Practice Tests

for

Andersen and Taylor's

Sociology: Understanding a Diverse Society
Fourth Edition

Practice Tests

for

Andersen and Taylor's

Sociology: Understanding a Diverse Society
Fourth Edition

Paulina X. Ruf
University of Tampa

THOMSON
WADSWORTH

© 2006 Thomson Wadsworth, a part of The Thomson Corporation. Thomson, the Star logo, and Wadsworth are trademarks used herein under license.

ALL RIGHTS RESERVED. No part of this work covered by the copyright hereon may be reproduced or used in any form or by any means—graphic, electronic, or mechanical, including photocopying, recording, taping, Web distribution, information storage and retrieval systems, or in any other manner—without the written permission of the publisher.

Printed in the United States of America
1 2 3 4 5 6 7 09 08 07 06 05

Printer: Darby Printing

ISBN 0-495-00065-5
Cover image: Passionate Fantasy by Hessam Abrishami

For more information about our products, contact us at:
Thomson Learning Academic Resource Center
1-800-423-0563

For permission to use material from this text or product, submit a request online at
http://www.thomsonrights.com.
Any additional questions about permissions can be submitted by email to thomsonrights@thomson.com.

Thomson Higher Education
10 Davis Drive
Belmont, CA 94002-3098
USA

Asia (including India)
Thomson Learning
5 Shenton Way
#01-01 UIC Building
Singapore 068808

Australia/New Zealand
Thomson Learning Australia
102 Dodds Street
Southbank, Victoria 3006
Australia

Canada
Thomson Nelson
1120 Birchmount Road
Toronto, Ontario M1K 5G4
Canada

UK/Europe/Middle East/Africa
Thomson Learning
High Holborn House
50–51 Bedford Road
London WC1R 4LR
United Kingdom

Latin America
Thomson Learning
Seneca, 53
Colonia Polanco
11560 Mexico
D.F. Mexico

Spain (including Portugal)
Thomson Paraninfo
Calle Magallanes, 25
28015 Madrid, Spain

TABLE OF CONTENTS

	Preface	
Chapter 1	Developing a Sociological Perspective	1
Chapter 2	Doing Sociological Research	11
Chapter 3	Culture	21
Chapter 4	Socialization	32
Chapter 5	Social Interaction and Social Structure	42
Chapter 6	Groups and Organizations	52
Chapter 7	Deviance	63
Chapter 8	Crime and Criminal Justice	74
Chapter 9	Social Class and Social Stratification	85
Chapter 10	Global Stratification	96
Chapter 11	Race and Ethnicity	108
Chapter 12	Gender	119
Chapter 13	Sexuality	130
Chapter 14	Age and Aging	141
Chapter 15	Families	151
Chapter 16	Education	162
Chapter 17	Religion	173
Chapter 18	Economy and Work	183
Chapter 19	Politics and Government	194
Chapter 20	Health Care	205
Chapter 21	Population, Urbanization, and the Environment	216
Chapter 22	Collective Behavior and Social Movements	227
Chapter 23	Social Change in Global Perspective	238

PREFACE

Welcome to the Practice Tests for the Fourth Edition of *Sociology: Understanding a Diverse Society* by Margaret L. Andersen and Howard F. Taylor. This booklet was designed to help you test and apply your knowledge of the chapter concepts as you read the textbook and prepare for examinations. Used in conjunction with the Study Guide, this booklet can be a valuable tool to reinforce your understanding of sociology at the introductory level.

Each chapter of the Practice Tests includes 50 Multiple Choice questions, 20 True/False questions, and 4 Short Answer questions, as well as an answer key. Use the page references in the answer key to revisit the text and review the chapter material for any question(s) that you may have answered incorrectly.

CHAPTER ONE

DEVELOPING A SOCIOLOGICAL PERSPECTIVE

Multiple Choice:

1. Sociology is defined as a:

   A. social science that investigates such social institutions as the economy, politics, economic systems, and families
   B. scientific study of human behavior in society
   C. discipline which has some characteristics in common with psychology, economics, and anthropology
   D. all of the above

2. _____ analyzes individual behavior.

   A. Sociology
   B. Anthropology
   C. Psychology
   D. none of the above

3. _____ is the study of human cultures.

   A. Psychology
   B. Political Science
   C. Anthropology
   D. Economics

4. The ability to identify the societal patterns that influence individual and group life is called:

   A. the sociological imagination
   B. social structure
   C. interpersonal dynamics
   D. a natural attitude

5. _____ affect large numbers of people and originate in the institutional arrangements and history of a society.

   A. Troubles
   B. Social structure
   C. Issues
   D. Social troubles

Chapter 1

6. The sociological imagination is the ability to:

    A. see that problems can be solved by changing the character of the individual
    B. distinguish between personal issues and public troubles
    C. see the connection between self and immediate relationships
    D. make a distinction between troubles and issues

7. Personal problems that are based on events or emotions in one's life are called:

    A. troubles
    B. manifest functions
    C. issues
    D. latent functions

8. Unemployment is a(n) _____ when it is caused by fundamental large-scale changes in the economy.

    A. trouble
    B. issue
    C. personal issue
    D. private matter

9. A big difference between sociology and common sense is that:

    A. sociologists do not utilize common sense
    B. common sense is a science
    C. sociology is not a science
    D. sociology is an empirical discipline

10. Peter Berger's term 'debunking' refers to:

    A. how sociologists view the world
    B. looking beyond the facades of everyday life to patterns and processes
    C. a weird Chinese cultural practice
    D. the face level of meaning and understanding

11. As part of the debunking function, sociology helps us to recognize the wide range of behaviors which occur in school beyond education, including:

    A. the formation of social cliques and their influence on students' self-concepts
    B. issues of gender equity in classrooms
    C. the effects of social class on the quality of available public education
    D. all of the above

12. Debunking requires stepping back from the taken-for-granted assumptions provided in everyday life - a process called:

    A. myth acceptance
    B. facade acceptance
    C. critical distance
    D. role play

13. The role of critical distance in developing a sociological imagination is well explained by sociologist:

    A. Georg Simmel
    B. Jesse Jackson
    C. Ella Baker
    D. Martin Luther King, Jr.

14. The sociologist who explored the role of the stranger within a group, who may provide unique insights into its structures and behaviors is:

    A. Herbert Spencer
    B. Max Weber
    C. Emile Durkheim
    D. Georg Simmel

15. Sociologists sometimes study and reveal very controversial topics and problems in society. These societal problems are referred to as:

    A. unsettling facts
    B. the pulse of society
    C. "what comes around, goes around"
    D. societal factors

16. The term _____ refers to the variety of group experiences that result from the social structure of society.

    A. societal significance
    B. pluralism
    C. diversity
    D. generation X

17. Some of the strongest diversity influences on the outcomes of people's lives are:

    A. money, sex, and power
    B. class, race, and gender
    C. politics and government policy
    D. music, art, and history

18. Society in global perspective refers to:

    A. the social and economic system of any one society is intertwined with other societies
    B. Disney's "It's A Small World After All"
    C. the philosophy of isolationism
    D. the economic dominance of the United States

Chapter 1

19. The principle that human reason can successfully direct social action for the improvement of society is:

    A. the Enlightenment
    B. positivism
    C. humanitarianism
    D. postmodernism

20. Which of the following social thinkers wrote *Democracy in America*?

    A. Auguste Comte
    B. Harriet Martineau
    C. Alexis de Tocqueville
    D. Emile Durkheim

21. The female sociologist who attacked slavery and the subservience of women in the United States was:

    A. Jane Addams
    B. Harriet Martineau
    C. Isabel Eaton
    D. Jane Austen

22. Which of the following sociologists wrote *How to Observe Manners and Morals*?

    A. Jane Addams
    B. Harriet Martineau
    C. Isabel Eaton
    D. Jane Austen

23. Emile Durkheim's research and theory, which explored social solidarity, was the basis for the _____ theoretical perspective.

    A. functionalist
    B. conflict
    C. symbolic interaction
    D. rational choice

24. According to Durkheim, the proper subject of sociology is the social patterns that are external to individuals, such as customs and values. These patterns are called:

    A. capitalism
    B. latent functions
    C. manifest functions
    D. social facts

4

25. Emile Durkheim believed that the sociologist's task is to study:

    A. social actions
    B. the transformative power of history
    C. social facts
    D. capitalism

26. Durkheim's phrase for people bonding together in society is called:

    A. social facts
    B. verstehen
    C. social solidarity
    D. sui generis

27. According to Durkheim, society is _____ to individuals.

    A. internal
    B. bilateral
    C. opposite
    D. external

28. Karl Marx envisioned the crux of sociology being about:

    A. economics
    B. people
    C. social causes
    D. labor movements

29. According to Karl Marx, the character of conflict is shaped directly and profoundly by:

    A. means of production
    B. solidarity
    C. social action
    D. sociological imagination

30. Which of the following statements is true, according to Karl Marx?

    A. The capitalist class owns the means of production
    B. The working class sell their labor in exchange for investments in the means of production
    C. Profit for the working class is produced through exploitation of the lower class
    D. all of the above

31. Max Weber recognized that society has three basic, interrelated dimensions, namely:

    A. race, class, and gender
    B. political, economic, and cultural
    C. upper, middle, and lower
    D. economic, educational, and religious

Chapter 1

32. Max Weber's concept of _____ refers to an understanding of social behavior from the point of view of those engaged in it.

    A.   social relativity
    B.   solidarity
    C.   sociological imagination
    D.   verstehen

33. Weber maintained that people do things in a context. and use their interpretive abilities to understand and give meaning to their lives. This is called:

    A.   verstehen
    B.   organic metaphor
    C.   social measure
    D.   social action

34. Because Herbert Spencer had faith that evolution always took a benign course toward perfection, he advocated a(n) _____ approach to social change.

    A.   active
    B.   intensive
    C.   cooperative
    D.   laissez-faire

35. _____ is the use of sociological research and theory in solving real human problems.

    A.   Organic metaphor
    B.   Applied sociology
    C.   Social action
    D.   Social Telesis

36. The Chicago School of American sociology:

    A.   evidenced an interest in how society shaped the mind and identities of individuals
    B.   supported a laissez-faire approach to social change
    C.   used social settings as laboratories for human research
    D.   both A and C are correct

37. Early American female sociologists:

    A.   were often excluded from university positions
    B.   often entered the lower-paying field of social work
    C.   were often active in the settlement house movement
    D.   all of the above

38. W. E. B. Du Bois was preoccupied with:

    A.   the course and consequences of conflict
    B.   empirical community studies
    C.   social facts
    D.   the ties that bind people to one another

39. Which of the following is NOT a term found in functional analysis?

    A. manifest function
    B. latent function
    C. misfunction
    D. dysfunction

40. Robert Merton realized that social practices have consequences that are not always immediately apparent nor necessarily the same as their stated purposes. The unintended consequences of behavior are called the:

    A. sui generis
    B. manifest functions
    C. latent functions
    D. covert functions

41. Karl Marx is an example of a famous:

    A. functional theorist
    B. conflict theorist
    C. symbolic interactionist
    D. communist

42. The theory which emphasizes the role of coercion and power in society is:

    A. functionalism
    B. conflict theory
    C. rational choice theory
    D. exchange theory

43. Which of the following statements best reflects conflict theory?

    A. Individuals occupy fixed social roles
    B. Individuals and society are interdependent
    C. Individuals are subordinate to society
    D. Individuals' roles contribute to social order

44. Symbolic interactionists believe that during interaction, the parties involved:

    A. make interpretations and these interpretations form the social bond
    B. immediately respond to their surroundings and to each other's actions
    C. do not have a shared system of symbols
    D. usually communicate effectively even if they do not share the same symbols

45. Which theory considers society to be socially constructed through human interpretation?

    A. Conflict Theory
    B. Symbolic Interactionist Theory
    C. Feminist Theory

Chapter 1

46. Which theory analyzes the status of women in society with the purpose of using that knowledge to better women's lives?

    A. Conflict Theory
    B. Symbolic Interactionist Theory
    C. Feminist Theory
    D. Postmodernist Theory

47. _____ argues that the behavior of individuals is determined by the rewards or punishments they receive in a day-to-day interaction with others.

    A. Postmodern Theory
    B. Rational Choice Theory
    C. Exchange Theory
    D. Conflict Theory

48. What theory is used by political scientists and economists to emphasize that the choices human beings make are guided by reason?

    A. Rational Choice Theory
    B. Post Modern Theory
    C. Symbolic Interactionist Theory
    D. Exchange Theory

49. The theory that argues that society is reflected in the words and images that people use to represent behavior and ideas is called:

    A. the feminist theory
    B. postmodernism
    C. exchange theory
    D. post functionalism

50. Postmodernism is based on the idea that society is not an objective thing, but rather is found in the _____ that people use to represent behavior and ideas.

    A. rational choices
    B. social facts
    C. societal ideations
    D. discourses

True/False Questions:

T   F   1.   Sociology is the systematic study of human history.

T   F   2.   Social institutions are established and organized systems of social behavior with a recognized purpose

T   F   3.   Karl Marx wrote the classic book *The Sociological Imagination*.

Developing a Sociological Perspective

T   F   4.   Troubles affect large numbers of people and have their origins in the institutional arrangements and history of a society.

T   F   5.   Debunking refers to dethroning the current social elite or bourgeoisie.

T   F   6.   Marginal people are those who share the dominant culture to some extent but are blocked from full participation.

T   F   7.   Race, class, and gender are the only sources of diversity in the United States.

T   F   8.   Positivism is a very important system of thought for the field of sociology in that it keys on accurate observation and description.

T   F   9.   Alexis de Tocqueville called sociology the "Queen of the Sciences."

T   F   10.   Harriet Martineau wrote a very successful book entitled *Society in America*.

T   F   11.   Emile Durkheim was intrigued by the bonds that link the members of a group.

T   F   12.   The pivotal economic concept in Karl Marx's theory was 'social facts.'

T   F   13.   Weber's concept of verstehen relates to behavior that people give meaning to.

T   F   14.   Social Darwinism holds to the "survival of the fittest" as the driving force of social evolution.

T   F   15.   Applied sociology is the use of sociological research and theory in solving real human problems.

T   F   16.   Chicago School of sociological thought developed much of their thinking and social application from Karl Marx.

T   F   17.   W. E. B. Du Bois wrote a famous community study called *The Philadelphia Negro*.

T   F   18.   Talcott Parsons saw functionalism as expressing the disconnectedness of society.

T   F   19.   In conflict theory, social order is maintained by domination and power.

T   F   20.   The individual and society are interdependent in symbolic interactionist theory.

Short Answer Questions:

1. What is the "sociological imagination"?
2. What is "humanitarianism" and how is it reflected in sociology?
3. Compare and contrast between "microsociology" and "macrosociology"?
4. Compare and contrast between functionalism, conflict theory, and symbolic interactionism.

Chapter 1

ANSWERS FOR CHAPTER ONE

Multiple Choice Questions:

1. D (p. 2)
2. C (p. 3)
3. C (p. 4)
4. A (p. 5)
5. C (p. 5)
6. B (p. 5)
7. A (p. 5)
8. B (p. 5)
9. D (p. 7)
10. B (p. 7)
11. D (p. 7)
12. C (p. 8)
13. A (p. 8)
14. D (p. 8)
15. A (p. 9)
16. C (p. 11)
17. B (p. 11)
18. A (p. 12)
19. C (p. 14)
20. C (p. 14)
21. B (p. 14)
22. B (p. 14)
23. A (p. 15)
24. D (p. 15)
25. C (p. 15)
26. C (p. 15)
27. D (p. 15)
28. A (p. 16)
29. A (p. 16)
30. D (p. 16)
31. B (p. 16)
32. D (pp. 16-17)
33. D (p. 17)
34. D (p. 17)
35. B (p. 17)
36. D (p. 17)
37. D (p. 18)
38. B (p. 19)
39. C (pp. 20-21)
40. C (p. 21)
41. B (p. 21)
42. B (p. 21)
43. C (p. 21)
44. A (p. 22)
45. B (p. 22)
46. C (p. 22)
47. C (p. 23)
48. A (p. 23)
49. B (p. 23)
50. D (p. 23)

True/False Questions:

1. F (p. 2)
2. T (p. 4)
3. F (p. 5)
4. F (p. 5)
5. F (p. 7)
6. T (p. 8)
7. F (p. 11)
8. T (p. 14)
9. F (p. 14)
10. T (p. 14)
11. T (p. 15)
12. F (p. 15)
13. F (pp. 16-17)
14. T (p. 17)
15. T (p. 17)
16. F (p. 17)
17. T (p. 19)
18. F (p. 21)
19. T (p. 21)
20. T (p. 22)

Short Answer Questions:

1. See p. 5
2. See p. 14
3. See p. 20
4. See pp. 20-23

CHAPTER TWO

DOING SOCIOLOGICAL RESEARCH

Multiple Choice:

1. In his study "Sidewalk," Duneier was engaged in what is called:

    A. participant observation research
    B. evaluation research
    C. survey research
    D. secondary research

2. When sociologists do research they engage in a process of:

    A. deceit
    B. imagination
    C. speculation
    D. discovery

3. When we say that science is empirical, we mean that it is based:

    A. mainly on conjecture
    B. on objective experiences and data
    C. on abstractions and subjective impressions
    D. upon all of the above

4. The process of scientific thought by which sociologists derive general conclusions from specific observations is:

    A. empirical reasoning
    B. common sense reasoning
    C. deductive reasoning
    D. inductive reasoning

5. Sociological studies based on interpretive observations, not statistical analysis, are called:

    A. illogical studies
    B. quantitative studies
    C. qualitative studies
    D. synthetic studies

6. _____ studies provide data from which one can calculate, for example, average incomes and the percent employed.

    A. Quantitative
    B. Qualitative
    C. Synthetic

Chapter 2

7. A _____ is a characteristic that can have more than one value or score.

    A. constant
    B. invariable
    C. variable
    D. hypothesis

8. A _____ is any abstract characteristic or attribute that can be potentially measured.

    A. constant
    B. concept
    C. invariable
    D. hypothesis

9. Conditions or characteristics that can have more than one value are:

    A. generalizations
    B. primary data
    C. secondary data
    D. variables

10. The _____ of a measurement is the degree to which it accurately measures or reflects a concept.

    A. reliability
    B. validity
    C. elasticity
    D. primacy

11. The reliability of a measurement is:

    A. the degree to which it accurately measures or reflects a concept
    B. often insured by researchers using more than one indicator for a particular concept
    C. the ability of a measurement to give the same result with repeated administrations
    D. both A and C are correct

12. _____ refers to the original material gathered by sociologists.

    A. Primary data
    B. Secondary data
    C. Tertiary data
    D. Personal data

13. For the General Social Survey, a sample of about _____ Americans is surveyed each year.

    A. 1,000
    B. 1,500
    C. 2,000
    D. 2,500

Doing Sociological Research

14. The National Opinion Research Center, which conducts the General Social Survey, is located at:

    A. the University of Chicago
    B. Harvard University
    C. the University of Michigan
    D. Yale University

15. Which of the following is considered a classic example of serendipity?

    A. Harvard studies
    B. Haworth studies
    C. Humphrey studies
    D. Hawthorne studies

16. Which of the following is a commonly used survey method of research?

    A. a questionnaire
    B. an interview
    C. a telephone poll
    D. All of the above are survey methods of research

17. In _____ questions, people must reply from a list of possible answers.

    A. qualitative
    B. open-ended
    C. closed-ended
    D. none of the above

18. A method in which the researcher becomes part of the group he/she is studying is called:

    A. participant observation
    B. survey
    C. content analysis
    D. secondary analysis

19. _____ permit the study of a large number of variables and results can be generalized to a larger population if sampling is accurate.

    A. Evaluation research
    B. Surveys
    C. Historical research
    D. Controlled experiments

20. _____ evaluates the actual outcomes of a program or strategy and often has direct policy application.

    A. Historical research
    B. Surveys
    C. Evaluation research
    D. Controlled experiments

Chapter 2

21. _____ focus on only two or three variables and allows the study of cause and effect.

    A. Controlled experiments
    B. Surveys
    C. Historical research
    D. Evaluation research

22. *Street Corner Society*, which documents one of the first participant observation studies, was written by:

    A. Judith Rollins
    B. Peter Berger
    C. William Foote Whyte
    D. C. Wright Mills

23. To conduct a controlled experiment, two groups are created. An experimental group and the:

    A. dependent group
    B. independent group
    C. control group
    D. equalized group

24. *Street Corner Society* is a classic work by sociologists:

    A. Judith Rollins
    B. William Foote Whyte
    C. C. Wright Mills
    D. Georg Simmel

25. A method that represents a way of using cultural artifacts to measure what people write, say, see, and hear, is called:

    A. field research
    B. participant observation
    C. secondary analysis
    D. content analysis

26. Research which examines social themes over time, by investigating the data in archives, including government records, private diaries, and church records, is called:

    A. comparative research
    B. evaluation research
    C. historical research
    D. policy research

27. The likelihood that a specific behavior or event will occur is:

    A. prediction
    B. correlation
    C. probability
    D. possibility

28. Evaluation research can be:

    A. qualitative
    B. quantitative
    C. either qualitative or quantitative
    D. none of the above

29. Closely allied with evaluation research is:

    A. market research
    B. historical research
    C. comparative research
    D. experimental research

30. A _____ is a relatively large collection of people that a research studies and about which generalizations are made.

    A. population
    B. sample
    C. organized group
    D. biased group

31. A _____ is any subset of a population.

    A. subpopulation
    B. sample
    C. organized group
    D. biased group

32. If a researcher were to interview a sample of people in an Irish pub, the research would likely be:

    A. as random a sample as any other
    B. representative of those who live in that geographical area
    C. overrepresentative of people who go to Irish pubs
    D. all of the above

33. The _____ is simply the value or score that appears most frequently in a set of data.

    A. correlate
    B. median
    C. mean
    D. mode

Chapter 2

34. The statistical tool used by researchers which is not skewed or distorted by extreme scores at either end is the:

   A. mean
   B. median
   C. mode
   D. percentage

35. Adding a list of 10 numbers and dividing by 10 would give one the _____ of the total.

   A. mode
   B. median
   C. rate
   D. mean

36. _____ is a widely used technique for analyzing the patters of association between pairs of variables like income and education.

   A. Prediction
   B. Probability
   C. Causation
   D. Correlation

37. The _____ of a correlation is simply how closely or tightly the variables are correlated or associated.

   A. reliability
   B. elasticity
   C. direction
   D. strength

38. _____ is a widely used method for analyzing data, where the data are broken down into subsets for comparison.

   A. Multiple-tabulation
   B. Cross-tabulation
   C. Nominal-tabulation
   D. Variable-tabulation

39. Correlation is not proof of:

   A. balance
   B. validity
   C. reliability
   D. causation

Doing Sociological Research

40. When a researcher uses statistics that stretch beyond the given sample population it is called:

    A. fudging
    B. overgeneralizing
    C. faking data
    D. using data selectively

41. When a researcher creates elements in an experiment that favor certain outcomes it is called:

    A. overgeneralizing
    B. fudging
    C. building in bias
    D. faking data

42. Perhaps one of the worse misuses of statistics is:

    A. fudging
    B. building in bias
    C. faking data
    D. overgeneralizing

43. When a researcher willfully uses only certain portions or aspects of an experiment it is called:

    A. using data selectively
    B. building in bias
    C. faking data
    D. fudging

44. Sociological researchers have an obligation to make their research:

    A. as objective and value-free as possible
    B. free to all interested parties
    C. free of political bias
    D. adhere to Max Weber's standards

45. _____, like empathy, is a way of sharing in the experiences of other people and an essential part of doing good research.

    A. *Sui generis*
    B. *Verstehen*
    C. *Tabula rasa*
    D. *Legalese*

46. Conducting a _____ session after an experiment means that the deception of its participants is temporary.

    A. logistical
    B. strategic
    C. debriefing
    D. revealing

Chapter 2

47. The U.S. government itself authorized the Tuskegee Syphilis Study to be continued until the early:

    A. 1970s
    B. 1960s
    C. 1950s
    D. 1940s

48. The professional code of ethics is clear that if a research subject is at risk of physical, mental, or legal harm, the subject must be:

    A. paid for their time
    B. informed of rights and responsibilities
    C. admitted to an institution
    D. referred to a counselor

49. _____ did her research on Black women domestic workers, she posed as a maid and did not tell her employers that she was a Ph.D. student writing her dissertation.

    A. Judith Asch
    B. Judith Humphrey
    C. Judith Hawthorne
    D. Judith Rollins

50. The purpose of the _____ Syphilis Study was to examine the effects of "untreated syphilis in the male negro."

    A. Nuremberg
    B. Topeka
    C. Tuskegee
    D. Macon

True/False Questions:

T   F   1.   Both qualitative and quantitative studies are NOT empirical.

T   F   2.   Sociological knowledge is not the same as philosophy or personal belief.

T   F   3.   While deductive reasoning is heavily based on the importance of observation and logical analysis, inductive reasoning is not.

T   F   4.   Quantitative research is research that is somewhat less structured and focuses on the question being asked.

T   F   5.   Qualitative research is research that only uses statistical methods.

T   F   6.   A concept is any abstract characteristic or attribute that can be potentially measured.

T   F   7. Knowing that they are being studied may cause people to change their behavior is a phenomenon called the *Hawthorne Effect*.

T   F   8. Data analysis can only be processed by statistical methods.

T   F   9. When a researcher stumbles across an unexpected finding it is called a serendipitous finding.

T   F   10. Generalization is the ability to draw conclusions from specific data and being able to apply them to a broader population.

T   F   11. Participant observation represents a type of quantitative research in sociology.

T   F   12. Controlled experiments are flexible and spontaneous ways of collecting data and are especially used for determining a pattern of cause and effect.

T   F   13. Content analysis is a way of measuring via such cultural artifacts as to what people write, say, see, and hear.

T   F   14. Evaluation research assesses the effect of policies and programs on people in society.

T   F   15. A probability is the likelihood that a specific behavior or event will NOT occur.

T   F   16. A population is a relatively large collection of people (or other units) that a researcher studies and about which generalizations are made.

T   F   17. A random sample is usually bias and does not offer everyone in a given population an equal chance.

T   F   18. A spurious correlation exists when there is no meaningful causal connection between apparently associated effects.

T   F   19. Cross-tabulation is a method for analyzing sociological data in which the data is broken down into subsets for comparison.

T   F   20. The professional code of ethics for sociologists makes a vague reference to whether a subject must be informed of the rights and responsibilities of both him/herself and of the researcher.

Short Answer Questions:

1. Compare and contrast between "deductive" and "inductive" reasoning.
2. What is the *Hawthorne Effect* and how does it impact social research?
3. Compare and contrast between evaluation research and market research.
4. List and discuss at least three of the statistical mistakes discussed in the text.

Chapter 2

ANSWERS FOR CHAPTER TWO

Multiple Choice Questions:

1.	A (p. 27)		50.	C (p. 48)
2.	D (p. 28)			
3.	B (p. 28)		**True/False Questions:**	
4.	D (p. 29)			
5.	C (p. 30)		1.	F (p. 28)
6.	A (p. 30)		2.	T (p. 28)
7.	C (p. 31)		3.	F (pp. 28-29)
8.	B (p. 31)		4.	F (p. 30)
9.	D (p. 31)		5.	F (p. 30)
10.	B (p. 32)		6.	T (p. 31)
11.	C (p. 32)		7.	T (p. 32)
12.	A (p. 33)		8.	F (p. 33)
13.	B (p. 33)		9.	T (p. 34)
14.	A (p. 33)		10.	T (p. 35)
15.	D (p. 34)???		11.	F (p. 37)
16.	D (p. 36)		12.	F (p. 38)
17.	C (p. 36)		13.	T (p. 40)
18.	A (p. 37)		14.	T (p. 41)
19.	B (p. 37, Table 2.1)		15.	F (p. 41)
20.	C (p. 37, Table 2.1)		16.	T (p. 42)
21.	A (p. 37, Table 2.1)		17.	F (p. 42)
22.	C (p. 38)		18.	T (p. 43)
23.	C (p. 38)		19.	T (p. 43)
24.	B (p. 38)		20.	F (p. 48)
25.	D (p. 40)			
26.	C (p. 40)		**Short Answer Questions:**	
27.	C (p. 41)			
28.	D (p. 41)		1.	See pp. 28-29
29.	A (p. 41)		2.	See p. 32
30.	A (p. 42)		3.	See p. 41
31.	B (p. 42)		4.	See p. 44
32.	C (p. 42)			
33.	D (p. 43)			
34.	B (p. 43)			
35.	D (p. 43)			
36.	D (p. 43)			
37.	D (p. 43)			
38.	B (p. 43)			
39.	D (p. 44)			
40.	B (p. 44)			
41.	C (p. 44)			
42.	C (p. 44)			
43.	A (p. 44)			
44.	A (p. 45)			
45.	B (p. 47)			
46.	C (p. 48)			
47.	A (p. 48)			
48.	B (p. 48)			
49.	D (p. 48)			

CHAPTER THREE

CULTURE

Multiple Choice:

1. Culture:

    A. is the complex system of meaning and behavior that defines the way of life for a given group or society
    B. includes ways of thinking as well as patterns of behavior
    C. is learned by humans through socialization
    D. all of the above

2. According to the text, the Tchikrin associate hair with:

    A. family ancestry
    B. physical strength
    C. evil spirits
    D. sexual power

3. The norms, laws, customs, ideas, and beliefs of a group of people are its:

    A. popular culture
    B. cultural capital
    C. non-material culture
    D. material culture

4. In any society, culture defines what is perceived as:

    A. beautiful and ugly
    B. right and wrong
    C. good and bad
    D. all of the above

5. The objects created by a culture are called:

    A. material culture
    B. nonmaterial culture
    C. substantial culture
    D. metaphysical culture

6. According to the text, _____ people are raised to be quiet and not outspoken.

    A. African American
    B. Native American
    C. Italian American

Chapter 3

7. The meaning of symbols is:

    A. dependent on the culture in which they appear
    B. separate and apart from the emotional attachments that guide human behavior
    C. inherent in themselves
    D. the same, regardless of the context in which they exist

8. The idea that something can be understood and judged only in relationship to the cultural context in which it appears is:

    A. cultural hegemony
    B. global culture
    C. cultural relativism
    D. ethnocentrism

9. The protests that have developed over symbols such as _____ are indicative of the enormous influence of cultural symbols.

    A. sports mascots
    B. the Confederate flag
    C. both A and B are true
    D. none of the above are true

10. British ethologist _____ lived for many years among the chimpanzees in Tanzania, Africa.

    A. Margaret Mead
    B. Beatrice Allen
    C. Michelle Koko
    D. Jane Goodall

11. Language:

    A. makes possible the formation of culture among humans.
    B. is a set of symbols and rules which, when put together in a meaningful way, provides a complex communication system.
    C. is fluid and dynamic.
    D. all of the above

12. The Sapir-Whorf hypothesis:

    A. argues that language determines other aspects of culture since it provides the categories through which social reality is defined and constructed
    B. argues that people who speak different languages have the same perception of reality
    C. argues that there us a weak link between the concept of time and culture
    D. adheres to all of the above

13. The theory that language determines other aspects of culture since language provides the categories through which social reality is defined and constructed is called the:

    A. Miller-Mallow hypothesis
    B. Griffiths-Gaab hypothesis
    C. Sapir-Whorf hypothesis
    D. Mahir-Jasim hypothesis

14. Alfred Blumenthal used the label Caucasian to refer to people from the Caucuses of:

    A. Russia
    B. Prussia
    C. Spain
    D. Great Britain

15. The significance of language in culture is particularly apparent in how patterns of _____ are reflected in language.

    A. race
    B. gender
    C. class
    D. all of the above

16. Throughout the period of _____ segregation in the American South, Black men, regardless of their age, were routinely referred to as "boy" by Whites.

    A. Jim Crow
    B. William Graham Sumner
    C. Sapir-Whorf
    D. none of the above

17. In the _____, "Black American" replaced the term "Negro" because the civil rights and Black Power movements inspired Black pride and the importance of self-naming.

    A. 1940s
    B. 1950s
    C. 1960s
    D. 1970s

18. The way we greet each in the United States is an example of:

    A. sanctions
    B. laws
    C. folkways
    D. mores

Chapter 3

19. William Graham Sumner referred to such behaviors as fashion, food preparation, and etiquette as:

    A. mores
    B. values
    C. folkways
    D. sanctions

20. Strict norms that control moral and ethical behavior are:

    A. values
    B. mores
    C. folkways
    D. sanctions

21. Which of the following statements about mores is false?

    A. Mores control moral and ethical behavior
    B. Mores are often formalized into laws
    C. Mores are generally loosely defined
    D. Mores are stricter norms than folkways

22. _____ are strict norms that control moral and ethical behavior.

    A. Mores
    B. Folkways
    C. Beliefs
    D. Norms

23. Which of the following is an example of a more?

    A. cheating in your sociology exam
    B. driving under the influence of illicit substances
    C. legal and religious sanctions against killing others
    D. all of the above

24. What are mechanisms of social control that enforce norms?

    A. folkways
    B. mores
    C. social sanctions
    D. beliefs

25. Ethnomethodology is a technique used for studying human interaction by:

    A. manipulating back-stage behavior
    B. focusing on front-stage behavior only
    C. deliberately disrupting social norms and observing how individuals respond
    D. none of the above

26. Beliefs are:

   A. expectations about what is considered appropriate behavior
   B. shared ideas held collectively by people within a given culture
   C. the universal ways in which humans develop thoughts
   D. unrelated to cultural norms and values

27. Which of the following statements about beliefs is true?

   A. Beliefs are the basis for many of a culture's norms and values
   B. Men wearing pants and not skirts is an example of a belief
   C. Beliefs are unrelated to a culture's norms and values
   D. Beliefs are related to a culture's values, but not to its norms

28. _____ theorists interpret beliefs as potentially competing world views.

   A. Functionalist
   B. Symbolic interaction
   C. Conflict
   D. none of the above

29. Values are:

   A. the written set of guidelines that define rewards and punishments in a society
   B. established cultural beliefs
   C. able to be categorized as folkways and mores
   D. the abstract standards in a society or group that define ideal principles and provide a general outline for behavior

30. The practice known as *potlatch*, was developed by the:

   A. veblen
   B. settlers
   C. Navajo
   D. Kwakiutl

31. What percentage of people living in the U.S. are foreign-born?

   A. 23
   B. 32
   C. 16
   D. 11

32. Which of the following is considered an indigenous musical art form that originated in the United States?

   A. Salsa
   B. Tango
   C. Polka

Chapter 3

33. _____ has roots in the musical traditions of the slave community and African cultures.

   A. Flamenco
   B. Jazz
   C. Cha-cha
   D. Rumba

34. Groups sensitive to the interests of minority cultures have criticized the dominant culture for being based so exclusively on:

   A. White cultural traditions
   B. male cultural traditions
   C. European cultural traditions
   D. all of the above

35. Prohibiting a group from expressing their culture may drive the culture underground, resulting in a:

   A. culture of fear
   B. culture of solidarity
   C. culture of diffusion
   D. culture of resistance

36. According to the text, President _____ directed "all Chinese, Japanese, and Korean children to attend the "Oriental School."

   A. John F. Kennedy
   B. Theodore Roosevelt
   C. Franklin Pierce
   D. George Washington

37. Like other _____, militia groups have unique modes of dress and a distinctive lifestyle.

   A. subcultures
   B. countercultures
   C. anti-cultures
   D. post-cultures

38. The habit of seeing things only from the point of view of one's own group is called:

   A. cultural relativism
   B. xenocentrism
   C. ethnocentrism
   D. multiculturalism

39. According to the text, _____ and the deep ethnic conflicts associated with it have produced many of the wars of contemporary world affairs.

   A. heterosexism
   B. androcentrism
   C. nationalism
   D. none of the above

40. A concept antithetical to ethnocentrism is _____, referring to modes of thinking that view society through the plural experiences of its diverse membership.

   A. multiculturalism
   B. ethnocentrism
   C. androcentrism
   D. xenocentrism

41. What percentage of homes in the U.S. have at least one television?

   A. 95
   B. 92
   C. 89
   D. 84

42. Television is ever-present in our lives that there are now homes called "_____" households, where television is on most of the time.

   A. constant television
   B. TV-maniac
   C. couch potato
   D. digital

43. In _____ percent of all U.S. households, the television is on most of the time.

   A. 72
   B. 62
   C. 52
   D. 42

44. According to the text, in the early _____, there was a more Afrocentric ideal of beauty with darker skin, Afro hairdos, and African clothing.

   A. 1950s
   B. 1960s
   C. 1970s
   D. 1980s

Chapter 3

45. The theory that contends that the mass media reflect the values of the general population is known as the:

    A. reflection hypothesis
    B. media effects theory
    C. globalization process
    D. media as mirror principle

46. In his book *Bowling Alone*, _____ argues that there has been a recent decline in civic engagement.

    A. Jane Addams
    B. Max Weber
    C. Robert Putnam
    D. C. Wright Mills

47. The concentration of cultural power, or the pervasive and excessive influence of one culture throughout society, is:

    A. cultural capital
    B. cultural diffusion
    C. cultural hegemony
    D. culture shock

48. The cultural resources that are socially designated as being worthy and that give advantages to groups possessing them is:

    A. cultural capital
    B. cultural diffusion
    C. cultural hegemony
    D. the dominant culture

49. The feeling of disorientation that can occur when one encounters a new or rapidly changing cultural situation is:

    A. cultural lag
    B. culture shock
    C. cultural diffusion
    D. cultural hegemony

50. Cultural changes can be caused by:

    A. cultural lag
    B. technological innovations
    C. a dominant group imposing a new culture on a society
    D. cultural diffusion

True/False Questions:

T   F   1.   Culture includes patterns of behavior, but not ways of thinking.

Culture

T   F   2.   Cultural relativism is the idea that something can be understood and judged only in relationship to the cultural context in which it appears.

T   F   3.   Sapir and Whorf think that language single-handedly dictates the perception of reality.

T   F   4.   The Sapir-Whorf hypothesis does not claim that language shapes how people define social reality.

T   F   5.   Folkways are the specific cultural expectations for how to behave in a given situation.

T   F   6.   Ethnomethodology is a technique for studying human interaction by deliberately disrupting social norms and observing how individuals respond.

T   F   7.   Beliefs are shared ideas held collectively by people within a given culture.

T   F   8.   The dominant culture is not necessarily the culture consisting of the most powerful group in society.

T   F   9.   Some subcultures retreat from dominant culture while others coexist.

T   F   10.   Countercultures are subcultures created as a reaction against the values of the dominant culture.

T   F   11.   Androcentrism is the habit of seeing things from the point of view of one's own group.

T   F   12.   The diffusion of a single culture throughout the world is referred to as ethnocentrism.

T   F   13.   Popular culture includes the beliefs, practices, and objects that are part of everyday traditions.

T   F   14.   Mass media refers to those channels of communication that are available to wide segments of the population.

T   F   15.   Television is the only form of popular culture that influences public consciousness about gender and race.

T   F   16.   The reflection hypothesis contends that the mass media reflects the values of the general population.

T   F   17.   Cultural hegemony refers to the cultural resources that are socially designated as being worthy and that give advantages to groups possessing such capital.

T   F   18.   Cultural lag refers to the delay in cultural adjustments to changing social conditions.

Chapter 3

T F 19. The feeling of disorientation that can occur when one encounters a new or rapidly changed cultural situation is called culture shock.

T F 20. Cultures changes in response to changed conditions in the society.

Short Answer Questions:

1. Compare and contrast between material and nonmaterial culture.
2. Define cultural relativism and provide an example.
3. Compare and contrast between implicit and explicit norms.
4. Define the reflection hypothesis.

ANSWERS FOR CHAPTER THREE

Multiple Choice Questions:

1. D (p.54)
2. D (p.54)
3. C (p.54)
4. D (p.54)
5. A (p.54)
6. B (p.56)
7. A (p.56)
8. C (p.57)
9. B (p.57)
10. D (p.58)
11. D (p.59)
12. A (p.59)
13. C (p.59)
14. A (p.60)
15. D (p.60)
16. A (p.61)
17. C (p.61)
18. C (p.62)
19. C (p.62)
20. B (p.62)
21. C (p.62)
22. A (p.62)
23. D (p.62)
24. C (p.62)
25. C (p.62)
26. B (p.63)
27. A (p.63)
28. C (p.63)
29. D (p.63)
30. D (p.63)
31. D (p.64)
32. D (p.64)
33. B (p.64)
34. D (p.65)
35. D (p.66)
36. B (p.66)
37. B (p.67)
38. C (p.67)
39. C (p.68)
40. A (p.68)
41. A (p.69)
42. A (p.70)
43. D (p.70)
44. C (p.71)
45. A (p.72)
46. C (p.73)
47. C (p.73)

50. D (p.76)

True/False Questions:

1. F (p.54)
2. T (p.57)
3. F (p.59)
4. F (p.59)
5. F (p.62)
6. T (p.62)
7. T (p.63)
8. F (p.64)
9. T (p.66)
10. T (p.67)
11. F (p.67)
12. F (p.68)
13. T (p.69)
14. T (p.69)
15. F (p.72)
16. T (p.72)
17. F (p.73)
18. T (p.75)
19. T (p.75)
20. T (p.75)

Short Answer Questions:

1. See p. 54
2. See p. 57
3. See p. 62
4. See p. 72

Chapter 4

CHAPTER FOUR

SOCIALIZATION

Multiple Choice:

1. The name given to individuals who have been raised in the complete absence of human contact is:

    A. social isolates
    B. autistic children
    C. attention-deficit children
    D. feral children

2. _____ refers to the process through which people learn the expectations of society.

    A. Education
    B. Symbolic interaction
    C. Social learning
    D. Socialization

3. It is through socialization that people absorb their culture in the form of:

    A. customs
    B. habits
    C. means of expression
    D. all of the above

4. "_____" occurs when behaviors and assumptions are learned so thoroughly that people no longer question them.

    A. Assimilation
    B. Cultural diffusion
    C. Internalization
    D. Cultural infusion

5. The expression _____ means humans are born as a "blank slate."

    A. *pluribus unum*
    B. *tabula rasa*
    C. *semper fidelis*
    D. *imprimatur*

6. _____ occurs when behaviors and assumptions are learned so thoroughly that people no longer question them.

    A. Internalization
    B. Absorption
    C. Individuation
    D. Socializing

7. Which of the following is true of socialization?

    A. It occurs primarily during childhood and the process is generally completed by age 21
    B. It is a life-long process
    C. It has little effect on one's capacity for role-taking
    D. It has minimal effect on one's self-esteem

8. The average young person, age 8-19, spends close to _____ hours per day immersed in media in various forms.

    A. five
    B. three
    C. nine
    D. seven

9. According to the text, every _____ seconds a high school student drops out.

    A. forty-three
    B. nine
    C. seventeen
    D. eleven

10. Analysts estimate that by age _____, the average child will have witnessed at least 18,000 simulated murders on television.

    A. twelve
    B. fourteen
    C. sixteen
    D. eighteen

11. According to the text, every _____ a young person under 25 dies from HIV infection.

    A. week
    B. minute
    C. hour
    D. day

12. Friends, fellow students, and coworkers are examples of:

    A. cliques
    B. in-groups
    C. secondary groups

Chapter 4

13. Michael Messner's research on men and sports reveals that:

    A. for most men, playing or watching sports is often the context for developing relationships with fathers, even when the father is absent or emotionally distant in other areas of life
    B. it was through sports relationships with male peers, more than anything else, that men's identity was shaped
    C. for many men, the athletic accomplishments of other family members created uncomfortable pressure to perform and compete
    D. All of the above were verified by Messner's research

14. In _____, when Title IX was passed, there were under 30,000 women total in college athletic programs, compared to 170,000 men.

    A. 1964
    B. 1972
    C. 1956
    D. 1978

15. The _____ consists of the informal and often subtle messages about social roles that are conveyed through classroom interaction and materials.

    A. latent curriculum
    B. secondary curriculum
    C. fulfilling prophecy
    D. hidden curriculum

16. Sigmund Freud provided the theoretical foundation for:

    A. social learning theory
    B. psychoanalytic theory
    C. social conflict
    D. the symbolic interaction perspective

17. Psychoanalytic theory originates in the work of:

    A. George H. Mead
    B. Nancy Chodorow
    C. Jean Piaget
    D. Sigmund Freud

18. According to the text, one of Freud's greatest contributions was the idea that:

    A. women were defined as psychologically better-adjusted than men
    B. it placed very little emphasis on biological drives and instead emphasized the social
    C. the unconscious mind shapes human behavior
    D. the human personality is learned

19. According to psychoanalytic theory:

    A. the conflict between the id and superego occurs in the subconscious mind
    B. "Freudian slips" reveal an underlying state of mind
    C. the forces that shape the self are located in the subconscious mind
    D. all of the above are true according to psychoanalytic theory

20. Your text identified four different theoretical perspectives that explain socialization. Which of the following is not one of those perspectives?

    A. social conflict theory
    B. psychoanalytic theory
    C. object relations theory
    D. symbolic interaction

21. Which of the following theories emphasizes the making and breaking of the bond with parents?

    A. social learning theory
    B. social conflict theory
    C. object relations theory
    D. rational exchange theory

22. Nancy Chodorow is most closely associated with:

    A. social learning theory
    B. object relations theory
    C. social conflict theory
    D. rational exchange theory

23. Social learning theory emphasizes:

    A. the societal context of socialization
    B. the importance of nature on socialization
    C. the belief that identity is the product of the unconscious
    D. all of the above

24. In the initial _____ stage, children experience the world only directly through their five senses.

    A. formal operational
    B. preoperational
    C. sensorimotor
    D. concrete operational

25. In the _____ stage, children are able to think abstractly and imagine alternatives to the reality in which they live.

    A. sensorimotor
    B. preoperational
    C. formal operational
    D. concrete operational

Chapter 4

26. Swiss psychologist Jean Piaget believed that:

    A. people were passive creatures who merely responded to stimuli in their environment
    B. while learning was crucial to socialization, imagination also had a critical role
    C. the importance of schema was overemphasized in theories of socialization
    D. there are really no distinctive stages of cognitive development

27. The initial state of development in Piaget's cognitive development theory is the _____ stage.

    A. concrete operational
    B. formal operational
    C. sensorimotor
    D. preoperational

28. According to Kohlberg, in the _____ stage, young children judge right and wrong in simple terms of obedience and punishment, based on their own needs and feelings.

    A. preconventional
    B. conventional
    C. postconventional
    D. unconventional

29. Symbolic interactionists use the term _____ to refer to a person's identity.

    A. superego
    B. personality
    C. ego
    D. self

30. According to symbolic interaction theory:

    A. the self ego emerges from tension between the id and the superego
    B. the self emerges through separating oneself from the primary caretaker
    C. the self is our concept of who we are and is formed in relationship to others
    D. the self is an interior bundle of drives, instincts, and motives

31. The looking-glass self is associated with:

    A. George Herbert Mead
    B. Erving Goffman
    C. Charles Horton Cooley
    D. Carol Gilligan

32. According to the text, the looking-glass self involves perception and:

    A. reflection
    B. game
    C. imitation
    D. effect

Socialization

33. Symbolic interaction theory suggests that children learn:

    A. as the unconscious mind shapes their behavior
    B. through taking the role of significant others
    C. only when they identify with the same sex parent
    D. all of the above

34. George Herbert Mead agreed with _____ that children are socialized by responding to other's attitudes toward them.

    A. Cooley
    B. Piaget
    C. Gilligan
    D. Goffman

35. George Herbert Mead saw childhood socialization as occurring in _____ stages.

    A. 5
    B. 3
    C. 4
    D. 2

36. According to Mead, the _____ stage is the second stage of childhood socialization.

    A. play
    B. imitation
    C. game
    D. reflection

37. Mead compared the lessons of the game stage to a:

    A. marathon
    B. basketball game
    C. baseball game
    D. school orientation

38. According to _____, social roles are learned in the family.

    A. conflict theory
    B. functionalism
    C. object relations theory
    D. psychoanalytic theory

39. According to _____, individuals learn social identities in the context of power relationships.

    A. functionalism
    B. conflict theory
    C. object relations theory
    D. psychoanalytic theory

Chapter 4

40. The socialization process is clearly _____ by factors such as class, race, gender, religion, regional background, sexual preference, age, and ethnicity.

    A. impervious
    B. unaffected
    C. patterned
    D. none of the above

41. _____ stated that the central task of adolescence is the formation of a consistent identity.

    A. Peter Berger
    B. George Herbert Mead
    C. Erik Erikson
    D. Jean Piaget

42. The learning of expectations associated with a role one expects to enter in the future is:

    A. resocialization
    B. career planning
    C. anticipatory socialization
    D. reality-based optimism

43. National surveys of young people today find that, by age _____, most young people have not achieved their occupational aspirations.

    A. 25
    B. 18
    C. 21
    D. 16

44. A ceremony or ritual that marks the passage of an individual from one role to another is:

    A. a rite of passage
    B. coming out
    C. anticipatory socialization
    D. resocialization

45. Graduation ceremonies, weddings, and bar mitzvahs are all examples of a:

    A. rite of transition
    B. rite of passage
    C. rite of separation
    D. rite of incorporation

46. The Latino celebration of quinceañera is a rite of passage that parallels a _____ in Anglo society.

    A. debutante's coming-out party
    B. retirement party
    C. bridal shower
    D. baby shower

47. _____ may involve physically and psychologically degrading new members with the aim of breaking down or redefining their old identity.

    A. Desocialization
    B. Resocialization
    C. Anticipatory socialization
    D. none of the above

48. Your text indicates that the case of John Walker Lindh, a U.S. citizen who joined the Taliban and was charged with conspiring to kill Americans, is one of extreme:

    A. desocialization
    B. anticipatory socialization
    C. conversion
    D. anti-socialization

49. Extreme examples of resocialization are seen in the phenomenon popularly called:

    A. desocialization
    B. anti-socialization
    C. psychosis
    D. brainwashing

50. The phenomenon of a captive identifying with the captor, which has been found among some hostages, prisoners of war, and battered women, is termed:

    A. psychosis
    B. psychic victimization
    C. the rape trauma syndrome
    D. the Stockholm syndrome

True/False Questions:

T F 1. Feral children usually overcome their developmental and language difficulties.

T F 2. Internalization occurs when behaviors and assumptions are learned so thoroughly that people no longer question them, but simply accept them as correct.

T F 3. Self-esteem has little to do with development of identity.

T F 4. Socialization agents are those individuals who pass on social expectations.

Chapter 4

T	F	5.	Religion is the first source of socialization for most individuals.
T	F	6.	There is no relationship to televised violence and aggressive behavior.
T	F	7.	Peers are enormously important in the socialization process of the individual.
T	F	8.	Messner's research on men and sports reveals the extent to which sports shape masculine identity.
T	F	9.	*Self-fulfilling prophecies* are positive appraisals, meaning that the expectations they create often become the basis for actual behavior.
T	F	10.	Social learning theory originates in the work of Sigmund Freud.
T	F	11.	Nancy Chodorow, a feminist sociologist, is renowned for her work in psychoanalytic theory.
T	F	12.	Social learning theory considers the formation of identity to be a learned response to social stimuli.
T	F	13.	Lawrence Kohlberg developed a theory of moral development based upon Piaget's stages of cognitive development.
T	F	14.	Carol Gilligan postulated the looking-glass self to explain how a person's conception of self arises through reflection about relationships to others.
T	F	15.	Charles Horton Cooley and George Herbert Mead were both sociologists at the University of Chicago in the early 1900s.
T	F	16.	According to George Herbert Mead, identity emerges from the roles one plays.
T	F	17.	Socialization begins with adolescence and ends with midlife crisis.
T	F	18.	Adult socialization is the process of learning new roles and expectations in adult life.
T	F	19.	A rite of passage is a ceremony or ritual that marks the passage of an individual from one role to another.
T	F	20.	An extreme example of resocialization can be seen in the phenomenon called the Stockholm syndrome.

Short Answer Questions:

1. List and discuss the three consequences of socialization discussed in your text.
2. What is the "hidden curriculum" and what role does it play in the socialization process?
3. Freud argued that the human psyche has three parts. List and discuss these three parts.
4. What does it mean to "take the role of the other"?

ANSWERS FOR CHAPTER FOUR

Multiple Choice Questions:

1. D (p.81)
2. D (p.83)
3. D (p.83)
4. C (p.83)
5. B (p.84)
6. A (p.84)
7. B (p.85)
8. D (p.86)
9. B (p.87)
10. D (p.87)
11. D (p.87)
12. D (p.88)
13. D (p.89)
14. B (p.90)
15. D (p.91)
16. B (p.92)
17. D (p.92)
18. C (p.92)
19. D (p.92)
20. A (p.92)
21. C (p.93)
22. B (p.93)
23. A (p.94)
24. C (p.94)
25. C (p.94)
26. B (p.94)
27. C (p.94)
28. A (p.94)
29. D (p.95)
30. C (p.95)
31. C (p.95)
32. D (p.95)
33. B (p.96)
34. A (p.96)
35. B (p.96)
36. A (p.96)
37. C (p.96)
38. B (p.97, Table 4.1)
39. B (p.97, Table 4.1)
40. C (p.99)
41. C (p.102)
42. C (p.103)
43. A (p.103)
44. A (p.104)
45. B (p.104)
46. A (p.105)
47. B (p.106)
48. C (p.106)
49. D (p.106)
50. D (p.107)

True/False Questions:

1. F (pp.81-82)
2. T (p.83)
3. F (p.85)
4. T (p.86)
5. F (p.86)
6. F (p.87)
7. T (p.88)
8. T (p.89)
9. F (pp.90-91)
10. F (p.92)
11. F (p.93)
12. T (p.94)
13. T (p.94)
14. F (pp. 94-95)
15. T (p.95)
16. T (p.96)
17. F (p.99)
18. T (p.102)
19. T (p.104)
20. T (p.107)

Short Answer Questions:

1. See pp. 85-86
2. See p. 91
3. See p. 92
4. See p. 96

Chapter 5

CHAPTER FIVE

SOCIAL INTERACTION AND SOCIAL STRUCTURE

Multiple Choice:

1. Human societies:

    A. have members that view themselves as distinct from other societies
    B. are systems of social interaction that include both culture and social organization
    C. have members which experience a high degree of interdependence
    D. all of the above

2. Behavior involving communication between two or more people is termed:

    A. a social institution
    B. impression management
    C. linguistics
    D. social interaction

3. _____ is meaningful behavior between two or more people.

    A. Social communication
    B. Social interaction
    C. Social action
    D. Social organization

4. Which of the following social thinkers saw society as an organism comprised of different parts that work together to create a unique whole?

    A. Karl Marx
    B. George Herbert Mead
    C. Emile Durkheim
    D. Erving Goffman

5. _____ described society as "*sui generis*."

    A. Emile Durkheim
    B. Auguste Comte
    C. Max Weber
    D. Karl Marx

6. The technique which sociologists use to investigate large patterns of social interaction to comprehend society as a whole is:

    A. gemeinschaft
    B. macroanalysis
    C. microanalysis
    D. proxemic communication

7. The _____ approach is utilized to study relatively small, less complex, and less differentiated patterns of social interaction.

   A. microlevel
   B. macrolevel
   C. middle-range
   D. organicist

8. _____ refers to the order established in social groups at any level.

   A. Social superstructure
   B. Social organization
   C. Social design
   D. Social infrastructure

9. Social organization:

   A. brings regularity and predictability to human behavior
   B. refers to people who are categorized together based on one or more shared characteristics
   C. describes the order established in social groups at any level
   D. both A and C are true

10. Sociologists would refer to all the people that watch the TV show *Fear Factor* as:

    A. a social group
    B. an audience
    C. a social constituency
    D. a social category

11. Which of the following statements about groups is true?

    A. A group is a collection of individuals who interact and communicate with each other
    B. Group members share goals and norms
    C. Group members possess a subjective awareness of themselves as a distinct social unit
    D. all of the above

12. A highly structured social grouping that forms to pursue a set of goals is a:

    A. social category
    B. gesellschaft
    C. formal organization
    D. social institution

13. Teenagers, truck drivers, and millionaires are all examples of:

    A. social categories
    B. social groups
    C. formal organizations
    D. informal organizations

Chapter 5

14. Your sex, race, and ethnicity are all examples of:

    A. acquired statuses
    B. assumed statuses
    C. achieved statuses
    D. ascribed statuses

15. Statuses which are automatically assigned to a person at birth, such as racial status, are termed:

    A. ascribed
    B. proxemic
    C. kinesic
    D. achieved

16. A lawyer, a medical doctor, and a college professor are all examples of:

    A. ascribed statuses
    B. assumed statuses
    C. achieved statuses
    D. acquired statuses

17. Role _____ occurs when the roles in one's role set clash with one another.

    A. inconsistency
    B. conflict
    C. strain
    D. breakdown

18. _____ are occupied; _____ are acted or "played."

    A. Statuses; roles
    B. Roles; statuses
    C. Strains; sets
    D. Sets; strains

19. _____ occurs when two or more roles are associated with contradictory expectations.

    A. Role set
    B. Role strain
    C. Role conflict
    D. Role subset

20. Hochschild's research illustrates _____, a condition that results from a single role that brings conflicting expectations.

    A. role strain
    B. role conflict
    C. role inconsistency
    D. role breakdown

Social Interaction and Social Structure

21. Which of the following sociologists coined the term "the second shift?"

    A. George Herbert Mead
    B. Peter Berger
    C. Arlie Hochschild
    D. Erving Goffman

22. W. I. Thomas argued that situations defined as real are real in their consequences. This theory is termed:

    A. the definition of the situation
    B. ethnomethodology
    C. impression management
    D. the social construction of reality

23. Which of the following social theorists is most closely associated with ethnomethodology?

    A. Goffman
    B. Berger
    C. Garfinkel
    D. Mills

24. Erving Goffman coined the term:

    A. "ethnomethodology"
    B. "impression management"
    C. "social construction"
    D. "background expectancies"

25. _____ can be seen as a type of *con game*.

    A. "Conflict resolution"
    B. "Beauty game"
    C. "Impression management"
    D. "Background expectancies"

26. _____ involves the stylist bridging the gap between those who seek beauty and those who define it.

    A. "Beauty myth"
    B. "Beauty game"
    C. "Beauty con"
    D. "Beauty work"

27. Goffman defines _____ as a spontaneous reaction to a sudden or transitory challenge to our identity.

    A. embarrassment
    B. conceit
    C. smugness
    D. none of the above

Chapter 5

28. The concept of "social profit" is most closely related to:

    A. impression management
    B. social exchange theory
    C. dramaturgy
    D. ethnomethodology

29. The "culture of calculation" is most closely associated with:

    A. Turkle
    B. Goffman
    C. Garfinkel
    D. Gimlin

30. The culture of _____ is reflected in the use of personal computers in that they provide experiences such as interaction in cyberspace.

    A. smugness
    B. cybernation
    C. simulation
    D. calculation

31. Kissing, hugging, and punching someone are all examples of:

    A. kinesic communication
    B. tactile communication
    C. paralinguistic communication
    D. none of the above

32. Sociologist Eli Anderson's book _____ notes how people use space to interact with strangers in the street.

    A. *The Street Code*
    B. *A Streetcar Name Desire*
    C. *Tally's Corner*
    D. *Streetwise*

33. One of the categories that social scientists classify nonverbal communication into is:

    A. tactile communication
    B. kinesic communication
    C. paralinguistic communication
    D. all of the above are categories nonverbal communication is divided into

34. When a baby cries, he/she is communicating:

    A. nothing
    B. kinesically
    C. verbally
    D. paralinguistically

Social Interaction and Social Structure

35. _____ means that emotions tend to 'leak out' even if a person tries to conceal them.

    A. "Information overflow"
    B. "Information overload"
    C. "Verbal leakage"
    D. "Nonverbal leakage"

36. _____ tend to stand much closer to each other than White middle-class Americans.

    A. Native Americans
    B. Asian Americans
    C. British Americans
    D. Hispanic Americans

37. According to the text, we tend to spend about _____ percent of our time with other people when doing all sorts of activities such as eating, watching television, studying, and working.

    A. 35
    B. 50
    C. 60
    D. 75

38. A _____ determinant of your attraction toward others is simply whether you live near them.

    A. faint
    B. immaterial
    C. weak
    D. strong

39. Our attraction to another is greatly affected by how frequently we see him/her even in pictures. This is called the _____ effect.

    A. overexposure
    B. mere exposure
    C. overexposed
    D. positive exposure

40. Research studies have found that teachers evaluate cute children of either gender as _____ than unattractive children with identical academic records.

    A. obtuse
    B. smarter
    C. scruffy
    D. dumb

41. Research studies have found that "dominant" people tend to be attracted to:

    A. "submissive" people
    B. other "dominant" people
    C. meek people
    D. none of the above

Chapter 5

42. _____ refers to the organized pattern of social relationships and social institutions that together comprise society.

    A. Cultural complex
    B. Material culture
    C. Social structure
    D. Bureaucracy

43. _____ see societal needs as universal, although societies do not perform them in the same way or by means of the same institutions.

    A. Conflict theorists
    B. Symbolic interactionists
    C. Functionalists
    D. Feminists

44. Which of the following theorists is associated with the concept of "collective consciousness?"

    A. Emile Durkheim
    B. Karl Marx
    C. Talcott Parsons
    D. E. T. Hall

45. "Contractual solidarity" is another name for _____ solidarity.

    A. mechanical
    B. organic
    C. collective
    D. conscious

46. In a _____ society, there is increasing importance on less intimate, more instrumental, secondary relationships. There is also a reduced sense of personal loyalty to the total society and a somewhat diminished role of the nuclear family.

    A. gemeinschaft
    B. foraging
    C. gesellschaft
    D. pastoral

47. Organic solidarity would typically be found in a:

    A. gesellschaft society
    B. relatively small society with limited division of labor
    C. gemeinschaft society
    D. both B and C are correct

Social Interaction and Social Structure

48. The economic process in which families become economically dependent on wages to support themselves, but work within the family is unpaid and increasingly devalued, is termed:

    A. the family-wage economy
    B. the feminization of poverty
    C. the agriculture-industry transition
    D. capitalistic development

49. Industrial societies tend to be highly productive economically with:

    A. with no serious social problems
    B. a large working class of industrial laborers
    C. major growth in rural areas
    D. families independent of the industrial system

50. In their well-received book, _____, sociologists Robert Bellah and colleagues argued that the individualistic orientation of people in the U.S. has created a society in which people find it difficult to sustain their commitments to others.

    A. *Theory of the Leisure Class*
    B. *Heart-wise Philosophy*
    C. *Habitat for Humanity*
    D. *Habits of the Heart*

True/False Questions:

T F 1. Macroanalysis refers to the technique which sociologists use to investigate patterns of social interaction that are relatively small, less complex, and less differentiated.

T F 2. All human social units are groups.

T F 3. A social category is comprised of all people nationwide or internationally who watch the same program or performance.

T F 4. Formal organizations are unstructured social groupings that are pursuing no definitive goals.

T F 5. Achieved statuses are those gained from the moment a person is born.

T F 6. Master statuses, since they override other identities for an individual, are often the basis for prejudice and stereotypes.

T F 7. Role conflict and role strain are experienced at the individual level, but their origins are societal, since they originate in the expectations rooted in specific roles.

T F 8. Ethnomethodology is a process by which people control how others will perceive them.

Chapter 5

T F 9. According to social exchange theory, persons present different "selves" to others in different settings by modifying behavior to willfully attempt to manipulate the other's impression of her or him.

T F 10. When two or more persons share a virtual reality experience via communication and interaction with each other, they are engaged in cyberspace interaction.

T F 11. Patterns of tactile communication are strongly influenced by gender. Research shows that in general females touch and hug to express sexual interest and males touch to express emotional support, such as male athletes who pat teammate's backs after unsuccessful plays.

T F 12. Kinesic communication is the component of communication that is conveyed by the pitch and loudness of the speaker's voice.

T F 13. Proxemic communication refers to the amount of space between interacting individuals.

T F 14. Although standards of attractiveness vary from culture to culture, considerable agreement exists within a culture about who is attractive.

T F 15. The mere exposure effect refers to those persons we become attracted to by seeing them more often than others.

T F 16. A social institution is an established and organized system of social behavior with a recognized purpose.

T F 17. Organic solidarity arises when individuals play similar roles within the society.

T F 18. In a gemeinschaft society, there is a sense of "we" feeling among members, a moderate division of labor, strong personal ties and family relationships, and a sense of personal loyalty.

T F 19. A postindustrial society is economically dependent on the production and distribution of services, information, and technology.

T F 20. The United States is a postindustrial society.

Short Answer Questions:

1. Compare and contrast between achieved and ascribed statuses.
2. Compare and contrast between ethnomethodology and impression management.
3. List and discuss the three types of nonverbal communication discussed in your text.
4. Describe the "mere exposure effect." Provide examples.

ANSWERS FOR CHAPTER FIVE

Multiple Choice Questions:

1. D (p.112)
2. D (p.112)
3. B (p.112)
4. C (p.112)
5. A (p.112)
6. B (p.112)
7. A (p.112)
8. B (p.112)
9. D (p.112)
10. B (p.113)
11. D (p.113)
12. C (p.113)
13. A (p.113)
14. D (p.113)
15. A (p.113)
16. C (p.113)
17. B (p.114)
18. A (p.114)
19. C (p.114)
20. A (p.115)
21. C (p.115)
22. A (p.116)
23. C (p.117)
24. B (p.117)
25. C (p.118)
26. D (p.118)
27. A (p.119)
28. B (p.120)
29. A (p.121)
30. C (p.122)
31. B (p.122)
32. D (p.122)
33. D (pp.122-123)
34. D (p.123)
35. D (p.123)
36. D (p.125)
37. D (p.126)
38. D (p.126)
39. B (p.127)
40. B (p.127)
41. B (p.128)
42. C (p.129)
43. C (p.129)
44. A (p.130)
45. B (p.130)
46. C (p.131)
47. A (p.131)
48. A (p.133)
49. B (p.133)
50. D (p.135)

True/False Questions:

1. F (p.112)
2. F (p.113)
3. F (p.113)
4. F (p.113)
5. F (p.113)
6. T (p.114)
7. T (pp.114-115)
8. F (p.117)
9. T (pp.119-120)
10. T (p.120)
11. F (p.123)
12. F (p.123)
13. T (pp.124-125)
14. T (p.127)
15. F (p.127)
16. T (p.128)
17. F (p.130)
18. T (p.131)
19. T (p.134)
20. F (p.134)

Short Answer Questions:

1. See p.113
2. See pp.117-118
3. See pp.122-125
4. See p.127

Chapter 6

CHAPTER SIX

GROUPS AND ORGANIZATIONS

Multiple Choice:

1. Which of the following statements about groups is (are) true?

    A. Juries are groups, and groups behave differently than individuals
    B. Predictions can be made about who will become the most influential jury member
    C. Jury verdicts correlate not just with the evidence but also with jury composition
    D. All of the above

2. A group:

    A. is a gathering of people
    B. is a social gathering, such as truck drivers and audiences
    C. may have a subjective awareness of "we," but this is not an essential characteristic of a group
    D. is a collection of individuals who interact with each other and share goals and norms

3. _____ was interested in discovering the effects of size on groups.

    A. Charles Horton Cooley
    B. Georg Simmel
    C. Emile Durkheim
    D. Karl Marx

4. "Expressive needs" are also called:

    A. task-oriented needs
    B. secondary needs
    C. intimacy needs
    D. socioemotional needs

5. Charles Horton Cooley introduced the concept of the:

    A. coalition
    B. triadic segregation
    C. primary group
    D. groupthink

Groups and Organizations

6. Cooley defined a group consisting of intimate, face-to-face interaction and relatively long-lasting relationships as a:

    A. primary group
    B. secondary group
    C. tertiary group
    D. formative group

7. Which of the following is an example of a secondary group?

    A. immediate family
    B. lifelong friends
    C. intimate partners
    D. co-workers

8. "Instrumental needs" are also called:

    A. task-oriented needs
    B. secondary needs
    C. intimacy needs
    D. socioemotional needs

9. According to the text, _____ are generalized versions of role models.

    A. secondary groups
    B. out-groups
    C. reference groups
    D. none of the above

10. _____ is the principle that we all make inferences about the personalities of others, such as concluding what another person is really like.

    A. Dispositional theory
    B. Attribution theory
    C. Deferential theory
    D. Risky shift theory

11. _____ noted that individuals commonly generate a significantly distorted perception of the motives and capabilities of other people's acts based on whether those people are in-group or out-group members.

    A. George Mead
    B. Thomas F. Pettigrew
    C. Robert Park
    D. Ernest Burgess

53

Chapter 6

12. Research has shown that identification with reference groups can have a strong effect on:

    A. self-evaluation
    B. self-esteem
    C. both A and B
    D. none of the above

13. Which of the following is LEAST likely to be considered an in-group?

    A. your family
    B. your sorority or fraternity
    C. your sports team
    D. your city

14. Members of the wealthy classes in the U.S. sometimes refer to one another as:

    A. "the gang"
    B. "the deserving ones"
    C. "people like us"
    D. "the lucky ones"

15. W. I. Thomas is credited with making the distinction between:

    A. in-groups and out-groups
    B. self-esteem and self-evaluation
    C. reference groups and generalized others
    D. none of the above

16. _____ refers to the perceived "true nature" of a person.

    A. "Attribute"
    B. "Serendipity"
    C. "Self-esteem"
    D. "Disposition"

17. The *Not-Me Syndrome* is associated with:

    A. Philip Zimbardo
    B. W. I. Thomas
    C. Charles H. Cooley
    D. Thomas F. Pettigrew

18. Which of the following individuals showed that even simple objective facts cannot withstand the distorting pressure of group influence?

    A. C. Wright Mills
    B. Solomon Asch
    C. Philip Zimbardo
    D. Peter Berger

Groups and Organizations

19. In the Milgram Obedience studies, the subjects acted as:

    A. teachers
    B. confederates
    C. learners
    D. scientists

20. Which of the following theorists wrote *Eichmann in Jerusalem*?

    A. Solomon Asch
    B. Stanley Milgram
    C. Hannah Arendt
    D. Philip Zimbardo

21. According to the text, the decision of the Naval High Command in 1941 not to prepare for attack on Pearl Harbor by Japan is an example of:

    A. "not-me syndrome"
    B. "small town" effect
    C. groupthink
    D. none of the above

22. In the spring of _____, it was revealed that American soldiers who were military police guards at a prison in Iraq had engaged in severe torture of Iraqi prisoners of war.

    A. 2002
    B. 2004
    C. 1992
    D. 1994

23. Which of the following social thinkers is associated with the 'groupthink' phenomenon?

    A. I. L. Janis
    B. Solomon Asch
    C. Philip Zimbardo
    D. Peter Berger

24. The tendency for group members to reach a consensus opinion, even if the decision is stupid, is:

    A. polarization shift
    B. status generalization
    C. group unanimity
    D. groupthink

25. Which of the following U.S. Presidents decided to increase the number of troops in Vietnam in 1967?

    A. Richard M. Nixon
    B. John F. Kennedy
    C. Lyndon B. Johnson
    D. Harry Truman

Chapter 6

26. "Polarization shift" is another name for:

   A. "intrapersonal anomie"
   B. "group dissonance"
   C. "risky shift"
   D. "not-me syndrome"

27. Outbreaks of groupthink usually:

   A. provide an illusion of vulnerability
   B. encourage dissenting opinions
   C. provide an illusion of unanimity
   D. none of the above

28. Risky shift was first observed by:

   A. I. L. Janis
   B. James Stoner
   C. Solomon Asch
   D. Stanley Milgram

29. The tendency for groups to weigh risk differently than individuals is called:

   A. risky shift
   B. triadic segregation
   C. polarization shift
   D. both A and C are correct

30. Which of the following is not an example of a formal organization?

   A. political parties
   B. work organizations
   C. the family
   D. schools

31. _____ is the sense that one's self has merged with a group.

   A. Deindividuation
   B. Anomie
   C. Panic
   D. Social disorganization

32. Which of the following is one of the three classic types of organizations identified by Etzioni?

   A. coercive organizations
   B. normative organizations
   C. utilitarian organizations
   D. all of the above

Groups and Organizations

33. Which of the following is an example of a normative organization?

    A. Kiwanis clubs
    B. political parties
    C. religious organizations
    D. all of the above

34. A _____ organization is one in which membership is largely involuntary.

    A. normative
    B. utilitarian
    C. coercive
    D. affiliative

35. The NAACP was founded in 1909 by _____ and the National Urban League.

    A. Solomon Asch
    B. B'Nai Brith
    C. C. Wright Mills
    D. W. E. B. Du Bois

36. Which of the following theorists is referred to coercive organizations as total institutions?

    A. James Stoner
    B. Erving Goffman
    C. Solomon Asch
    D. Stanley Milgram

37. Which of the following is an example of coercive organizations?

    A. B'Nai Brith
    B. the NAACP
    C. La Raza Unida
    D. mental hospitals

38. _____ include two populations: the inmates and the staff.

    A. Unitarian institutions
    B. Normative institutions
    C. Utilization institutions
    D. Total institutions

39. Large organizations that individuals join for specific purposes (e.g. for monetary rewards) are called _____ organizations.

    A. normative
    B. voluntary
    C. coercive
    D. utilitarian

Chapter 6

40. As various organizations – like mental hospitals – become privatized, some utilitarian organizations may also be:

   A. normative organizations
   B. functional organizations
   C. voluntary organizations
   D. coercive organizations

41. The early sociological theorist _____ analyzed the classic characteristics of the bureaucracy.

   A. Emile Durkheim
   B. Max Weber
   C. Talcott Parsons
   D. Karl Marx

42. Sociologist _____ notes that many modern bureaucracies have hierarchical authority structures and an elaborate division of labor.

   A. Paula Giddings
   B. James Stoner
   C. Talcott Parsons
   D. Charles Perrow

43. _____ is characterized by the individual becoming psychologically separated from the organization and its goals. This process may result in increased turnover, tardiness, absenteeism, and overall dissatisfaction with the organization.

   A. Ritualism
   B. Alienation
   C. Risky shift
   D. Groupthink

44. Your authors suggest that the primary organizational principle that lies behind McDonaldization is:

   A. an emphasis on efficiency
   B. calculability
   C. control, whereby peoples behavior, both customers and workers alike, is reduced to a series of machine-like actions
   D. predictability

45. The primary organization principle that lies behind McDonaldization is:

   A. flexibility
   B. predictability
   C. control
   D. calculability

58

Groups and Organizations

46. Drawing partly on early sociological theorists _____ and Emile Durkheim, Ritzer argued that "these cathedrals of consumption" have created a spectacular and simulated world.

    A. Max Weber
    B. Ferdinand Tönnies
    C. Talcott Parsons
    D. Peter Berger

47. Which theory of management is based on the premise that by nature, people have a desire to work, and also to be creative and to take on responsibility if individual self-expression is permitted?

    A. Theory X
    B. Self-actualization
    C. Theory Z
    D. Self-management

48. This theory predicts that an organization can operate efficiently and produce if the self-direction of the individual is permitted expression is called:

    A. self-actualization
    B. Theory X
    C. Theory Y
    D. Theory Z

49. The _____ effect means that women and minorities may be promoted but only to a certain level.

    A. slow track
    B. down escalator
    C. glass ceiling
    D. mere ceiling

50. Max Weber argued that certain functions, called _____, characterize bureaucracies and contribute to its overall unity.

    A. malfunctions
    B. dysfunctions
    C. eufunctions
    D. none of the above

True/False Questions:

T   F   1. The difference between a dyad and a triad creates no significantly difference in group dynamics.

T   F   2. A triad is a group consisting of exactly three people.

T   F   3. A dyad is an unstable social grouping, whereas triads are relatively stable.

Chapter 6

T	F	4.	Primary groups are those to which you may or may not belong, but that you use as a standard for evaluating your values, attitudes, and behaviors.
T	F	5.	A social network is a set of dispositional attributions set by others under certain conditions.
T	F	6.	The conviction that one is impervious to social influence results in what social psychologist Max Weber called the Not-Me Syndrome.
T	F	7.	Milgram, in his study of conformity, found that gender, class, race, and ethnic differences had no significant effects on compliance rates.
T	F	8.	Groupthink is the tendency for group members to reach a consensus opinion only when the decision is most effective for the organization..
T	F	9.	While organizations tend to be persistent, they also can be tools for innovation.
T	F	10.	Gender, class, race, and ethnicity play a role in who joins what voluntary organizations.
T	F	11.	Coercive organizations are characterized by membership that is largely voluntary.
T	F	12.	The early sociological theorist Karl Marx analyzed the classic characteristics of the bureaucracy.
T	F	13.	Sexual harassment can become an aspect of the informal culture of the workplace.
T	F	14.	Rigid adherence to rules can produce a slavish following of them, whether or not it accomplishes the purpose for which the rules were originally designed.
T	F	15.	Predictability in the McDonaldization process refers to the emphasis on the quantitative aspects of products sold, such as size, cost, and time of production, rather than the quality of products.
T	F	16.	In recent years, U.S. competition with Japan has spawned interest in Japanese styles of management.
T	F	17.	There are no definitive patterns of race, gender, and age discrimination in organizations.
T	F	18.	Kanter demonstrated in her research how the hierarchical structure of the bureaucracy negatively affects both minorities and women.
T	F	19.	Social class, in addition to race and gender, plays a part in determining people's place within formal organizations.
T	F	20.	Max Weber argued that *proletariat functions* characterize bureaucracies and contribute to the overall unity of the bureaucracy.

Short Answer Questions:

1. Compare and contrast between primary and secondary groups.
2. Describe the Asch Conformity Experiment.
3. Compare and contrast between groupthink and risky shift.
4. List and discuss the six characteristics of a bureaucracy presented in the text.

Chapter 6

ANSWERS FOR CHAPTER 6

Multiple Choice Questions:

1. D (p.140)
2. D (p.140)
3. B (p.140)
4. D (p.141)
5. C (p.141)
6. A (p.141)
7. D (p.141)
8. A (p.142)
9. C (p.142)
10. B (p.143)
11. B (p.143)
12. C (p.143)
13. D (p.143)
14. C (p.143)
15. A (p.143)
16. D (p.144)
17. A (p.145)
18. B (p.146)
19. A (p.146)
20. C (pp.147-148)
21. C (p.148)
22. B (p.148)
23. A (p.148)
24. D (p.148)
25. C (p.149)
26. C (p.149)
27. C (p.149)
28. B (p.149)
29. D (p.149)
30. C (p.150)
31. A (p.150)
32. D (p.151)
33. D (P.151)
34. C (p.151)
35. D (p.151)
36. B (p.151)
37. D (p.151)
38. D (p.151)
39. D (pp.151-152)
40. D (p.152)
41. B (p.152)
42. D (p.152)
43. B (p.155)
44. C (p.156)
45. C (p.156)
46. A (p.157)
47. B (pp.157-158)
48. A (pp.157-158)
49. C (p.158)
50. C (p.160)

True/False Questions:

1. F (p.140)
2. T (p.140)
3. F (p.141)
4. F (p.141)
5. F (p.144)
6. F (p.145)
7. T (p.147)
8. F (p.148)
9. T (p.150)
10. T (p.151)
11. F (p.151)
12. F (p.152)
13. T (p.153)
14. T (p.154)
15. F (p.156)
16. T (p.158)
17. F (p.158)
18. T (p.159)
19. T (p.159)
20. F (p.160)

Short Answer Questions:

1. See p.141
2. See p.146
3. See pp.148-150
4. See pp.152-153

CHAPTER SEVEN

DEVIANCE

Multiple Choice:

1. For the survivors of an airplane that crashed in the Andes Mountains in the early 1970s, _____ was generally accepted as something they had to do to survive.

    A. consuming snow
    B. gathering berries
    C. fishing
    D. cannibalism

2. _____ is more than simple nonconformity; it is behavior that departs significantly from social expectations.

    A. Kowtowing
    B. Convergence
    C. Junction
    D. Deviance

3. _____ deviance is behavior that breaks laws.

    A. Informal
    B. Formal
    C. Situational
    D. Convergent

4. Which of the following is an example of informal deviance?

    A. cheating on a test
    B. jaywalking
    C. body piercing
    D. larceny

5. Which of the following is NOT an example of formal deviance?

    A. theft
    B. rape
    C. jaywalking
    D. tattooing

Chapter 7

6. _____ is an example of informal deviance.

    A. Cutting in line
    B. Body piercing
    C. Tattooing
    D. All of the above

7. _____ is stressed by the sociological definition of deviance.

    A. Genetic background
    B. Individual behavior
    C. Social context
    D. none of the above

8. _____ considered deviance "functional" for society.

    A. Erving Goffman
    B. Emile Durkheim
    C. George H. Mead
    D. Karl Marx

9. According to the text, deviance tends to _____ society; by defining some forms of behavior as deviant, people are affirming the social norms of groups.

    A. stabilize
    B. weaken
    C. expose
    D. permeate

10. Sociologist _____ argues that the image of crack-addicted mothers as harming innocent babies has created a moral panic, in which low-income Black mothers are blamed for an "epidemic" of drug abuse.

    A. Stanley Milgram
    B. Drew Humphries
    C. C. Wright Mills
    D. Peter Berger

11. According to the text, by 2000 close to _____ the public thought that smoking should be banned in restaurants.

    A. all of
    B. a quarter
    C. a third
    D. half

12. The gay and lesbian movement has:

    A. not advocated new rights for lesbian women and gay men
    B. encouraged people to see gay and lesbian relationships as legitimate
    C. supported the public label of gays and lesbians as deviant
    D. supported all of the above

13. Deviance may violate social norms, but people:

    A. never recognize it as doing this
    B. do not always disapprove of the behavior
    C. rarely know what norms are being violated
    D. none of the above

14. Psychological explanations of deviance emphasize individual _____ factors as the underlying cause of deviant behavior.

    A. drives
    B. physiological
    C. upbringing
    D. personality

15. Biological explanations attribute deviance to:

    A. personality structures
    B. social environment
    C. presumed genetic difference
    D. evil spirits

16. According to Ferguson, school personnel think of the _____ as doing well in school.

    A. *Schoolboys*
    B. *Hood boys*
    C. *Gang bangers*
    D. *Lost boys*

17. Most deviance, to most sociologists, is not a pathological state, but an adaptation to:

    A. the social structures in which people live
    B. the circumstances under which people were diagnosed
    C. their individual personality structures
    D. none of the above

18. Which of the following perspectives supports the idea that deviance "creates social cohesion"?

    A. symbolic interactionism
    B. conflict theory
    C. functionalism
    D. feminist theory

Chapter 7

19. _____ is an important concept in Durkheim's studies of deviance.

    A. "Alienation"
    B. "Anomie"
    C. "Androgeny"
    D. "Anonymity"

20. Anomie, as defined by Durkheim, is frequently referred to as a state of:

    A. euphoria
    B. excessive social integration
    C. total detachment from society
    D. relative normlessness

21. Suicide on college campuses is often cited as an example of _____ suicide.

    A. anomic
    B. altruistic
    C. egoistic
    D. situational

22. Which of the following is an example of altruistic suicide?

    A. suicide bombers
    B. elderly man with no one left
    C. college student that is lonely
    D. a young woman with a history of sexual abuse

23. Durkheim's major point is that suicide is a significantly _____ phenomenon.

    A. individual
    B. personal
    C. social
    D. none of the above

24. Which of the following theorists is associated with structural strain theory?

    A. W. I. Thomas
    B. Travis Hirschi
    C. Robert Merton
    D. Robert Park

25. Structural strain theory most closely reflects:

    A. symbolic interactionism
    B. functionalism
    C. conflict theory
    D. feminist theory

Deviance

26. According to Merton, which of the following groups is most likely to experience structural strains?

    A. rich people
    B. the upper class
    C. the middle class
    D. poor people

27. _____ theory traces the origins of deviance to the tensions caused by the gap between cultural goals and the means people have to achieve these goals.

    A. Differential association
    B. Structural strain
    C. Structural association
    D. Social bond

28. According to the text, prostitution is a type of _____ deviance.

    A. retreatism
    B. ritualistic
    C. innovative
    D. none of the above

29. According to the text, a college student who suffers from bulimia is an illustration of _____ deviance.

    A. ritualistic
    B. retreatism
    C. innovative
    D. none of the above

30. According to the text, a homeless person illustrates _____ deviance.

    A. retreatism
    B. ritualistic
    C. innovative
    D. none of the above

31. Many right-wing extremist groups, such as the American Nazi Party and the Ku Klux Klan, are examples of _____ deviance.

    A. retreatism
    B. ritualistic
    C. innovative
    D. rebellion

Chapter 7

32. Which of the following is not an example of *retreatism deviance*?

   A. the homeless
   B. skinheads
   C. the severe alcoholic
   D. the hermit

33. Social control theory most closely reflects:

   A. symbolic interactionism
   B. conflict theory
   C. functionalism
   D. feminist theory

34. Social control theory is associated with the work of:

   A. Robert Merton
   B. Travis Hirschi
   C. Karl Marx
   D. Edwin Sutherland

35. Travis Hirschi developed _____ theory to explain the occurrence of deviance.

   A. social control
   B. structural strain
   C. differential association
   D. labeling

36. According to _____, the economic organization of capitalist societies produces deviance and crime.

   A. functionalism
   B. symbolic interactionism
   C. conflict theory
   D. none of the above

37. _____ is crime committed by the elite within the legitimate context of doing business.

   A. Capitalist crime
   B. CEO-deviance
   C. Blue-Collar crime
   D. Corporate crime

38. Which of the following theorists is associated with *white-collar crime*?

   A. Travis Hirschi
   B. Edwin Sutherland
   C. Robert Merton
   D. Emile Durkheim

39. _____ emphasizes the significance of social control in managing deviance and crime, in that controlling social deviance is one way that dominant groups control the behaviors of others.

   A. Conflict theory
   B. Functionalism
   C. Symbolic interactionism
   D. Social exchange theory

40. Symbolic interactionism emphasizes the _____ surrounding deviance.

   A. power relationships
   B. cohesion
   C. structural strain
   D. meanings

41. Differential association theory most closely reflects:

   A. functionalism
   B. feminist theory
   C. conflict theory
   D. symbolic interactionism

42. _____ theory interprets deviance as behavior one learns through interaction with others.

   A. Differential association
   B. Structural strain
   C. Structural association
   D. Social bond

43. Which of the following theories are microsociological theories?

   A. functionalism
   B. symbolic interaction theory
   C. conflict theory
   D. all of the above

44. Critics of _____ theory argue that this perspective tends to blame deviance on the values of particular groups.

   A. social control
   B. social bond
   C. structural strain
   D. differential association

Chapter 7

45. Another contribution of _____ theory is the understanding that deviance refers not just to something one does, but also to something one becomes.

   A. structural strain
   B. labeling theory
   C. social bond
   D. social control

46. _____ theory suggests that by recognizing mental illness, society also upholds normative values about more conforming behavior.

   A. Conflict
   B. Functionalist
   C. Symbolic interactionist
   D. Feminist

47. Of the following groups, which is least likely to be labeled mentally ill?

   A. White women
   B. Black men
   C. White men
   D. Black women

48. All of the following are examples of social stigmas, except:

   A. being physically disabled
   B. being disfigured
   C. being overweight
   D. all of the above are social stigmas

49. According to the text, alcohol is most likely to be used by those aged:

   A. 18-34
   B. 12-17
   C. 35-49
   D. 50-64

50. Just as _____ is shaping other dimensions of social life, so is it shaping deviant activities.

   A. globalization
   B. separatism
   C. nationalism
   D. localization

Deviance

True/False Questions:

T   F   1.   The definition of deviance remains constant over time and through cultures.

T   F   2.   People routinely engage in deviant acts, never thinking of themselves as deviant or having others label them as such.

T   F   3.   Some of the earliest attempts to explain deviance centered on psychological explanations.

T   F   4.   Biological explanations of deviance emphasize individual personality factors as the underlying cause of deviant behavior

T   F   5.   According to functionalism, deviance is learned behavior.

T   F   6.   Medicalizing deviance emphasizes the physical and/or genetic roots of deviant behavior.

T   F   7.   Anomie is defined as the condition that exists when social regulations in a society break down.

T   F   8.   Durkheim's structural strain theory traces the origins of deviance to the tensions caused by the gap between cultural goals and the means people have to reach them.

T   F   9.   Elite deviance is the term used to refer to the wrongdoing of wealthy and powerful individuals and organizations.

T   F   10. Social control agents are those who regulate and administer the response to deviance, such as police and mental health workers.

T   F   11. George Herbert Mead was among the first to develop a sociological perspective on social deviance.

T   F   12. Tertiary deviance is the actual violation of a norm or law.

T   F   13. Secondary deviance is the behavior that results from being labeled deviant.

T   F   14. A deviant career refers to the sequence of movements people make through a particular subculture of deviance.

T   F   15. Labeling theorists think that official rates of deviance reflect the actual incidence of deviance.

T   F   16. Some groups are organized around particular forms of social deviance and called deviant communities.

T   F   17. A master status is an attribute that is socially devalued and discredited.

Chapter 7

T F 18. A characteristic of a person that overrides all other features of the person's identity is called a master status.

T F 19. Stigmatized individuals are measured against a presumed norm and may be labeled, stereotyped, and discriminated against.

T F 20. Drug and alcohol use varies substantially by gender and race.

Short Answer Questions:

1. Compare and contrast between formal and informal deviance.
2. List and discuss the three types of suicide discussed by Durkheim.
3. Define what is meant by *white-collar crime*. Provide examples.
4. What are *deviant communities*?

ANSWERS FOR CHAPTER SEVEN

Multiple Choice Questions:

1. D (p.165)
2. D (p.166)
3. B (p.166)
4. C (p.166)
5. D (p.166)
6. A (p.166)
7. C (p.166)
8. B (p.167)
9. A (p.167)
10. B (p.168)
11. D (p.168)
12. B (p.168)
13. B (p.169)
14. D (p.169)
15. C (p.170)
16. A (p.170)
17. A (p.171)
18. C (p.171)
19. B (p.172)
20. D (p.172)
21. A (p.173)
22. A (p.173)
23. C (p.173)
24. C (p.173)
25. B (p.173)
26. D (p.174)
27. B (p.174)
28. C (p.174)
29. A (p.175)
30. A (p.175)
31. D (p.175)
32. B (p.175)
33. C (p.175)
34. B (p.175)
35. A (p.175)
36. C (p.177)
37. D (p.177)
38. B (p.177)
39. A (pp.177-178)
40. D (p.178)
41. D (p.178)
42. A (p.178)
43. B (p.178)
44. D (p.179)
45. B (p.180)
46. B (p.182)
47. C (p.183)
48. D (p.183)
49. A (p.185)
50. A (p.187)

True/False Questions:

1. F (p.166)
2. T (p.169)
3. F (p.170)
5. F (pp.170-171)
4. F (p.171)
6. T (p.171)
7. T (p.172)
8. F (p.174)
9. T (p.177)
10. T (p.178)
11. F (p.178)
12. F (p.180)
13. T (p.180)
14. T (p.180)
15. F (p.181)
16. T (p.181)
17. F (p.183)
18. T (p.183)
19. T (p.183)
20. T (p.185)

Short Answer Questions:

1. See p.166
2. See pp.172-173
3. See p.177
4. See p.181

Chapter 8

CHAPTER EIGHT

CRIME AND CRIMINAL JUSTICE

Multiple Choice:

1. Certain crimes, such as assault and robbery, are _____ to go on record if the person committing them is a person of color.

    A. less likely
    B. more likely
    C. unlikely
    D. none of the above

2. Criminologists include social scientists such as _____ who stress the societal causes of crime, and _____ who stress the personality-based causes of crime.

    A. psychologists; sociologists
    B. anthropologists; political scientists
    C. sociologists; psychologists
    D. none of the above

3. Which of the following theorists would assert that "crime may be functional to society"?

    A. a conflict theorist
    B. a functionalist
    C. a symbolic interactionist
    D. a feminist theorist

4. Which of the following theorists would suggest that disadvantaged groups are more likely to become criminals than those who are privileged?

    A. a symbolic interactionist
    B. a functionalist
    C. a conflict theorist
    D. a feminist theorist

5. In addition to the *Uniform Crime Reports*, a second major source of crime data are the:

    A. *Criminal Justice Newsletter*
    B. *Criminal Activity Annual Reports*
    C. *Crime Bureau Reports*
    D. *National Crime Victimization Surveys*

Crime and Criminal Justice

6. Since the 1990s, violent crimes have:

    A. decreased
    B. increased
    C. remained the same
    D. none of the above

7. The first kind of crime data is provided by:

    A. the FBI
    B. the police departments across the country
    C. the CIA
    D. the local newspapers

8. An especially notorious youth gang of the 1960s and 1970s in Chicago was called the:

    A. "Crips"
    B. "Bloods"
    C. "Blackstone Rangers"
    D. "Rukans"

9. Which of the following is the author of *The Gang* (1927)?

    A. Frederick M. Thrasher
    B. Travis Hirschi
    C. Edwin Sutherland
    D. Martin Sanchez Jankowski

10. Probation and parole officers are likely to describe juvenile girls in sexualized ways, particularly when the girl is:

    A. Native American
    B. Latina
    C. White, non-Latina
    D. African American

11. Murder, aggravated assault, forcible rape, and robbery are all examples of:

    A. power crimes
    B. victimless crimes
    C. property crimes
    D. personal crimes

12. Arson, larceny, and auto theft are all examples of:

    A. power crimes
    B. property crimes
    C. victimless crimes
    D. personal crimes

Chapter 8

13. The *Uniform Crime Reports* classify crimes into:

    A. property crimes
    B. victimless crimes
    C. personal crimes
    D. all the above

14. _____ crimes are motivated by various forms of bias, including those based on race, religion, sexual orientation, ethnic or national origin, or disability.

    A. Esteem
    B. Elite
    C. Malicious
    D. Hate

15. Presently, _____ states have enacted laws specifically directed against hate crimes as a particular category of crime.

    A. few
    B. all
    C. most
    D. an unknown number of

16. In _____, Martha Stewart was sentenced to five months in prison.

    A. 2004
    B. 2003
    C. 2002
    D. 2001

17. An example of elite or "white collar" crime would be:

    A. Theodore Kaczynski (for mail bombings)
    B. John Gotti (for organized crime)
    C. Ivan Boesky (for insider trading)
    D. none of the above

18. _____ crime and corporate crime are so highly organized, complex, and sophisticated that they take on the nature of social institutions.

    A. Organized
    B. Hate
    C. Personal
    D. none of the above

19. Corporate crime and deviance is wrongdoing that occurs within the context of a _____ organization or bureaucracy.

    A. informal
    B. formal
    C. unceremonious
    D. casual

20. According to the text, tax cheaters in business alone probably skim $_____ billion a year from the Internal Revenue Service.

    A. 50
    B. 100
    C. 250
    D. 500

21. Taken as a whole, the cost of corporate crime is almost _____ times the amount taken in bank robberies in a given year.

    A. 60
    B. 100
    C. 6000
    D. 1000

22. An example of organizational deviance, and possibly crime, is:

    A. WorldCom's accounting fraud
    B. Michael Milkmen's junk bond manipulation
    C. Ivan Boesky's insider trading
    D. LTC Oliver North's connection to the "Iran-Contra" scandal

23. Former Enron Vice President _____ has been labeled a "whistle-blower" and has found it difficult to find a job.

    A. Andrew S. Fastow
    B. Scott D. Sullivan
    C. Sherron Watkins
    D. David F. Myers

24. Sociologist _____ found a clear link between the likelihood of lethal violence and the socioeconomic conditions for Latinos in different cities.

    A. Martin Sanchez Jankowski
    B. Ramiro Martinez Jr.
    C. Ivan Boesky
    D. Frederick M. Thrasher

Chapter 8

25. The *Uniform Crime Reports* show a clear pattern of differential arrest along the lines of:

    A. race
    B. gender
    C. class
    D. all the above

26. Minorities constitute ____ percent of the U.S. population but more than one in three of those people are arrested for property crimes and almost one-half for violent crimes.

    A. 25
    B. 10
    C. 8
    D. 3

27. Which minority group has one of the lowest arrest rate for crimes?

    A. African Americans
    B. Asian Americans
    C. Hispanic Americans
    D. Russian Americans

28. According to the text, young _____ women are especially vulnerable to violent crime.

    A. White, non-Hispanic
    B. Hispanic
    C. Black
    D. Native American

29. According to the text, women's participation in crime has _____ in recent years.

    A. decreased
    B. dwindle
    C. increased
    D. none of the above

30. The Bureau of Justice Statistics finds that _____ percent of college women experience rape or attempted rape in a given college year.

    A. 13
    B. 3
    C. 30
    D. 23

31. The Bureau of Justice Statistics finds that _____ percent of college women report being stalked.

    A. 3
    B. 30
    C. 13
    D. 23

32. According to sociologist _____, women's fear of crime results from an ideology that depicts women as needing protection from men.

    A. Martin Sanchez Jankowski
    B. Ramiro Martinez Jr.
    C. Ivan Boesky
    D. Esther Madriz

33. For all women, victimization by _____ is probably their greatest fear.

    A. vandalism
    B. burglary
    C. robbery
    D. rape

34. More than _____ rapes are reported to the police annually.

    A. 500,000
    B. 400,000
    C. 100,000
    D. 200,000

35. "Driving while Black" is an example of:

    A. racial profiling
    B. police profiling
    C. not-me syndrome
    D. mere exposure effect

36. According to the text, annually at least _____ percent of all African Americans are *not* arrested.

    A. 80
    B. 70
    C. 90
    D. 60

37. At sentencing, Blacks and _____ are likely to get longer sentences than _____ for the same crimes, even when they have the same number of prior arrests and the same socioeconomic background.

    A. Whites; Latinos
    B. Asian; Latinos
    C. Asian; Native Americans
    D. Latinos; Whites

Chapter 8

38. The total cost to the nation of keeping people behind bars is approximately:

    A. $100 billion
    B. $150 billion
    C. $50 billion
    D. $200 billion

39. What minority group has the largest percentage on death row?

    A. Hispanic Americans
    B. African Americans
    C. Native Americans
    D. Asian Americans

40. Women comprise only _____ percent of all state and federal prisoners.

    A. 17
    B. 15
    C. 12
    D. 8

41. The typical woman in prison is a poor, young minority who dropped out of high school, is unmarried, and is:

    A. not a U.S. citizen
    B. asexual
    C. mentally ill
    D. the mother of two or more children

42. Of all women prisoners, about _____ have been victims of sexual abuse.

    A. half
    B. two-thirds
    C. a quarter
    D. none of the above

43. In general, prisons _____ seem to deter or rehabilitate offenders.

    A. never
    B. constantly
    C. always
    D. rarely

44. The Supreme Court decision in _____ (1965) gave married women the right to birth control.

    A. *Romer v. Evans*
    B. *Eisenstadt v. Baird*
    C. *Brown v. Board of Education of Topeka*
    D. *Griswold v. Connecticut*

45. In _____ (1996), the Supreme Court rule that states cannot pass laws that deprive gays and lesbians the equal protection of the law promised under the Fourteenth Amendment to the U.S. Constitution.

   A. *Eisenstadt v. Baird*
   B. *Romer v. Evans*
   C. *Brown v. Board of Education of Topeka*
   D. *Griswold v. Connecticut*

46. According to the text, about _____ percent of lawyers are women.

   A. 10
   B. 20
   C. 30
   D. 50

47. Decisions that critically affect the lives of the American people lie in the hands of a fairly _____ group of individuals.

   A. diverse
   B. heterogeneous
   C. homogeneous
   D. assorted

48. The role of all courts is to interpret and enforce:

   A. the law
   B. justice
   C. revenge
   D. morality

49. According to the text, the _____ defines terrorism as the unlawful use of force or violence against persons or property to intimidate or coerce a government or population in furtherance of political or social objectives.

   A. FBI
   B. international community
   C. United Nations
   D. none of the above

50. "Logic bombs" are utilized to attack:

   A. print media
   B. a computer system
   C. the postal service
   D. none of the above

Chapter 8

True/False Questions:

T F 1. Conflict theory has helped us understand how people learn to become criminals.

T F 2. Assault and robbery decreased quite significantly through the 1990s.

T F 3. The National Institute of Justice says that crime is increasing and hit an all-time high in 1999.

T F 4. It is uncommon for the gang member to think of her or his immediate subdivision of the gang as a family.

T F 5. The number of hate crimes that are reported has decreased in recent years.

T F 6. Women traditionally have been excluded from meaningful leadership roles in organized crime.

T F 7. The Soprano Effect is the extortion of money from legitimate small as well as large businesses on a regular basis.

T F 8. The structure of crime in the U.S. often takes on an organized character.

T F 9. Sociologists estimate that the costs of corporate crime may be as high as $200 billion every year.

T F 10. Those at the lowest ends of the socioeconomic scale are far more likely to be victims of violent crime.

T F 11. Native Americans have a relatively low rate of arrests for crimes.

T F 12. African Americans are generally less likely to be victimized by crime.

T F 13. People in the most disadvantaged groups are more likely to be defined and identified as criminal.

T F 14. The United States and Russia have the highest rates of incarceration in the world.

T F 15. Native Americans are overrepresented in prisons.

T F 16. The typical woman in prison is old and White, non-Hispanic.

T F 17. The rate of violence in private prisons is higher than in state facilities.

T F 18. Women's participation in crime has decreased in recent years.

T F 19. According to Reiman, the prison system in the U.S. is designed to train and socialize inmates into a career of crime.

T F 20. About 25 percent of lawyers are Hispanic.

Short Answer Questions:

1. Compare and contrast between personal, property and victimless crimes.
2. Compare and contrast between organized and corporate crime.
3. Define racial profiling. Provide examples.
4. Compare and contrast between *de jure* and *de facto* segregation.

Chapter 8

ANSWERS FOR CHAPTER EIGHT

Multiple Choice Questions:

1. B (p.191)
2. C (p.192)
3. B (p.192, Table 8.1)
4. C (p.192)
5. D (p.193)
6. A (p.193)
7. B (p.193)
8. C (p.194)
9. A (p.194)
10. B (p.194)
11. D (p.194)
12. B (p.194)
13. D (p.194)
14. D (p.195)
15. C (p.195)
16. A (p.195)
17. C (p.195)
18. A (p.196)
19. B (p.196)
20. A (p.196)
21. C (p.196)
22. A (p.197)
23. C (p.197)
24. B (p.198)
25. D (p.198)
26. A (p.198)
27. B (p.198)
28. C (p.200)
29. C (p.201)
30. B (p.201)
31. C (p.201)
32. D (p.201)
33. D (p.201)
34. D (p.201)
35. A (p.202)
36. C (p.203)
37. D (p.203)
38. B (p.203)
39. B (p.203)
40. D (p.204)
41. D (p.204)
42. B (p.205)
43. D (p.205)
44. D (p.206)
45. B (p.206)
46. C (p.206)
47. C (p.206)
48. A (p.206)
49. A (p.207)
50. B (p.207)

True/False Questions:

1. F (p.192)
2. T (p.193)
3. F (p.193)
4. F (p.194)
5. F (p.195)
6. T (p.196)
7. F (p.196)
8. T (p.196)
9. T (p.196)
10. T (p.198)
11. T (p.198)
12. F (p.199)
13. T (p.201)
14. T (p.203)
15. T (p.203)
16. F (p.204)
17. T (p.204)
18. F (p.204)
19. T (p.205)
20. F (p.206)

Short Answer Questions:

1. See pp.194-195
2. See pp. 196-197
3. See pp.202-203
4. See p.206

CHAPTER NINE

SOCIAL CLASS AND SOCIAL STRATIFICATION

Multiple Choice:

1. A relatively fixed, hierarchical arrangement in society by which groups have different access to resources, power, and perceived social wealth is termed:

    A. the ascribed status system
    B. the achieved status system
    C. apartheid
    D. social stratification

2. Black family income is _____ percent of White family income.

    A. 54
    B. 76
    C. 92
    D. 68

3. _____ percent of the population controls 38 percent of the total wealth in the nation.

    A. Ten
    B. One
    C. Five
    D. Fifteen

4. Estate systems of stratification are most common in:

    A. post-industrial societies
    B. hunting and gathering societies
    C. agricultural societies
    D. industrial societies

5. Jim Crow segregation in the American South is another example of a:

    A. indentured system
    B. class system
    C. estate system
    D. caste system

Chapter 9

6. Class systems are more open than _____ systems, because position does not depend strictly on birth.

    A. caste
    B. estate
    C. feudal
    D. none of the above

7. Old money families in New England are commonly called the:

    A. "Boston Kshityras"
    B. "Boston Caste"
    C. "Boston Brahmans"
    D. "Boston Untouchables"

8. Which of the following social theorists is associated with the term "life chances"?

    A. Karl Marx
    B. Max Weber
    C. Emile Durkheim
    D. Herbert Spencer

9. Content analyses of the media reveal that representations of welfare _____ themes of dependency.

    A. overemphasize
    B. deemphasize
    C. ignore
    D. disregard

10. Which of the following exemplify Weber's idea of "life chances"?

    A. the opportunity for possessing goods
    B. having an income
    C. having access to particular jobs
    D. all of the above are examples of life chances

11. Max Weber called to the opportunities that people have in common by virtue of belonging to a particular class:

    A. strata indicators
    B. life chances
    C. social attributes
    D. social anomalies

12. Class standing determines (or strongly influences):

    A. how well one is served by social institutions
    B. ones' political and social attitudes
    C. friendships that one develops
    D. all of the above

Social Class and Social Stratification

13. The _____ includes small business owners and managers whom you might think of as middle class.

    A. proletariat
    B. petty bourgeoisie
    C. lumpenproletariat
    D. none of the above

14. Capitalists extract profit by keeping the cost of _____ down.

    A. merchandise
    B. commodities
    C. labor
    D. none of the above

15. Weber saw three dimensions to stratification. Which of the following is not one of these dimensions?

    A. personality
    B. power
    C. class
    D. status

16. "Class," according to Weber, reflects the _____ dimension of stratification.

    A. social
    B. cultural
    C. political
    D. economic

17. Prestige is related to economic standing but may be independent of:

    A. income
    B. status
    C. social value
    D. none of the above

18. "Different parts of the social system complement one another and are held together through social consensus and cooperation." This statement most closely reflects:

    A. feminist theory
    B. conflict theory
    C. functionalism
    D. symbolic interactionism

Chapter 9

19. To explain stratification, _____ propose that the roles filled by the upper classes are essential for a cohesive and smoothly running society.

   A. conflict theorists
   B. functionalists
   C. symbolic interactionists
   D. social constructionists

20. Which of the following theorists analyzed the functions of poverty?

   A. Herbert Gans
   B. Randall Collins
   C. Max Weber
   D. Ralf Dahrendorf

21. "Different groups struggle over societal resources and compete for social advantage." This statement most closely reflects:

   A. functionalism
   B. symbolic interactionism
   C. conflict theory
   D. social exchange theory

22. "Beliefs about success and failure confirm status on those who succeed." This statement most closely reflects:

   A. functionalism
   B. symbolic interactionism
   C. conflict theory
   D. social exchange theory

23. In 2002, the median household income in the United States was:

   A. $52,409
   B. $42,409
   C. $32,409
   D. $22,409

24. The more education people think is needed for a given occupation, the more _____ people attribute to that job.

   A. "job prestige"
   B. "job cachet"
   C. "occupational prestige"
   D. "occupational cachet"

25. Educational attainment is typically measured as the:

    A. total years of informal education
    B. total years of formal education
    C. total years needed for completion of a high school diploma
    D. total years one attended school

26. To be on the 2003 list of the Forbes 400, your net worth had to be at least:

    A. $600 million
    B. $1 billion
    C. $900 million
    D. $100 million

27. According to the text, the four hundred richest Americans have a total net worth that exceeds the gross domestic product of the entire nation of:

    A. Britain
    B. the United States
    C. Japan
    D. China

28. Sullivan and colleagues found that bankruptcy is a _____-class phenomenon representing a cross section of people in this class.

    A. lower
    B. upper
    C. middle
    D. none of the above

29. Members of the upper class with newly acquired wealth are known as the:

    A. *nouveau riche*
    B. *rich trash*
    C. *dot com riche*
    D. *Forbes riche*

30. According to the text, _____ percent of the poor hold jobs.

    A. 44
    B. 57
    C. 38
    D. 21

31. The American dream of owning a home means many people are _____ - extended beyond their earning capability.

    A. "bankruptcy ready"
    B. "downward bound"
    C. "credit-driven"
    D. "mortgage poor"

Chapter 9

32. The middle class is also called _____ and includes managers, supervisors, and professional workers, such as doctors, lawyers, and professors.

   A. the lumpen-class
   B. the professional-managerial class
   C. the dynamic class
   D. none of the above

33. One thing that becomes clear with careful study of the U.S. class structure is the fact that:

   A. there is enormous class inequality in this society, and it is growing
   B. the importance of power is greatly overrated as a determining factor in the class status that one occupies
   C. this is basically a middle class society
   D. the gap between the rich and the poor is decreasing

34. Which of the following statements is true about wealth?

   A. Reliable data on the distribution of wealth is abundant and easily accessible
   B. Wealth is the amount of money brought into a household from wages, investments, income, and dividends
   C. Wealth is the monetary value of everything one actually owns
   D. The wealthiest 1 percent of the population own 57 percent of all the net wealth

35. In the U.S., the wealthiest 1 percent own _____ percent of all net worth.

   A. 73
   B. 61
   C. 52
   D. 38

36. According to the text, _____ percent of Americans have zero or negative net worth, usually because their debt exceeds their assets.

   A. 8
   B. 31
   C. 23
   D. 18

37. For every dollar of wealth held by White Americans, Black Americans have only:

   A. 76 cents
   B. 26 cents
   C. 36 cents
   D. 56 cents

Social Class and Social Stratification

38. The U.S. tax structure has distributed benefits:

    A. evenly
    B. unevenly
    C. fairly
    D. none of the above

39. Oliver and Shapiro's study of racial inequality in the distribution of wealth concludes that:

    A. a need exists for policies that promote development of all groups at the bottom of the social structure
    B. a need exists for a massive redistribution of wealth to reverse the historical advantage that some groups have had and now use to their cumulative benefit
    C. solid evidence exists that both race and class distinctions contribute to a perpetuation of inequality in the U.S. today
    D. all of the above are true

40. While most Americans are paying more in federal tax than ever before, _____ taxes since 1990 have fallen from 26 cents on the dollar to 20 cents.

    A. income
    B. sales
    C. state
    D. corporate

41. According to your text, the biggest gap in retirement income is between:

    A. men and women
    B. married and unmarried people
    C. native born and foreign born people
    D. none of the above

42. Researchers find that 46 percent of people in the United States identifies as:

    A. lower-class
    B. under-class
    C. upper-class
    D. middle-class

43. The classic sociologist who originated the term false consciousness was:

    A. Herbert Spencer
    B. Emile Durkheim
    C. Karl Marx
    D. August Comte

Chapter 9

44. Sociologists have found that the single most important determinant of where one sees oneself in the class system is:

   A. one's annual gross income
   B. their parents' socioeconomic status
   C. whether one does mental or manual labor
   D. how much they perceive their lives to be controlled by others

45. Which of the following is not true about social mobility in the United States today?

   A. The U.S. is a land of opportunity where anyone who works hard enough can get ahead
   B. Social mobility occurs in the U.S., but less often than the myth asserts and over shorter distances from one class to another
   C. Most people remain in their class of origin
   D. All of the above are myths about social mobility in the U.S. today

46. Since the 1950s, poverty has _____ in the United States.

   A. increased
   B. declined
   C. remained the same
   D. none of the above

47. A survey of twenty-seven cities has found that the homeless population is half:

   A. Mexican American
   B. African American
   C. White, non-Hispanic
   D. Native American

48. Reduction in federal support programs for the poor also contributes to the:

   A. graying of poverty
   B. feminization of poverty
   C. internationalization of poverty
   D. none of the above

49. The "culture of poverty" argument is associated with the work of:

   A. Herbert Gans
   B. Oscar Lewis
   C. Ralf Dahrendorf
   D. Randall Collins

50. TANF stipulates a lifetime limit of _____ years for people to receive aid and requires all welfare recipients to find work within two years.

   A. 5
   B. 3
   C. 2
   D. 4

Social Class and Social Stratification

True/False Questions:

T F 1. Social stratification is the process by which different statuses in any group are organized.

T F 2. In a caste system, one's place in the stratification system is an ascribed status.

T F 3. The system of apartheid in South Africa was a stark example of an estate system.

T F 4. Much of Marx's analysis boils down to the consequences of a system based on the pursuit of equality.

T F 5. Weber agreed with Marx that economic forces are the primary dimension of stratification.

T F 6. Weber used the term "status" to refer to the prestige dimension of stratification.

T F 7. As a functionalist, Herbert Gans held that poverty is necessary to sustain an overall social system.

T F 8. Conflict theorists argue that the consequences of inequality are positive for society.

T F 9. Status attainment is the process by which people end up in a given position in the stratification system.

T F 10. Occupational prestige refers to the subjective evaluation people give to jobs

T F 11. In the U.S. the largest group in the class system is the lower class.

T F 12. Conflict theorists tend to see the poor as being necessary for society to develop.

T F 13. Unlike income, the value of wealth tends to increase through investment.

T F 14. Factors such as age, ethnicity, race, gender, and national origin have a tremendous influence on the stratification of societies.

T F 15. The United States has been typically characterized as a class-conscious society.

T F 16. Karl Marx used the term false consciousness to describe the class consciousness of subordinate classes who had internalized the view of the dominant class.

T F 17. The class system in the U.S. is characterized as a closed class system since there is no chance for upward mobility.

T F 18. Homelessness has substantially increased over the past two decades.

Chapter 9

T   F   19.   The public stereotype that poverty is passed through generations is not well supported by the facts.

T   F   20.   Sociologists conclude that the so-called welfare trap is a matter of learned dependency.

Short Answer Questions:

1. Compare and contrast between social differentiation and social stratification.
2. List and discuss the three dimensions of stratification presented by Max Weber.
3. Compare and contrast between class and false consciousness.
4. List and discuss the two explanations of poverty presented in the text.

Social Class and Social Stratification

ANSWERS FOR CHAPTER NINE

Multiple Choice Questions:

1. D (p.212)
2. D (p.212)
3. B (p.212)
4. C (p.213)
5. D (p.214)
6. A (p.214)
7. C (p.214)
8. B (p.215)
9. A (p.215)
10. D (p.215)
11. B (p.215)
12. D (p.215)
13. B (p.216)
14. C (p.216)
15. A (p.216)
16. D (p.216)
17. A (p.217)
18. C (p.217)
19. B (p.217)
20. A (p.217)
21. C (p.218, Table 9.2)
22. A (p.218, Table 9.2)
23. B (p.220)
24. C (p.220)
25. B (p.220)
26. A (p.221)
27. D (p.221)
28. C (p.222)
29. A (p.222)
30. C (p.223)
31. D (p.223)
32. B (p.224)
33. A (p.225)
34. C (p.226)
35. D (p.226)
36. D (p.227)
37. B (p.227)
38. B (p.228)
39. D (p.228)
40. D (p.228)
41. B (p.231)
42. D (p.232)
43. C (p.232)
44. C (p.232)
45. A (p.233)
46. B (p.234)
47. B (p.235)
48. B (p.236)
49. B (p.237)
50. A (p.240)

True/False Questions:

1. F (p.212)
2. T (p.213)
3. F (p.213)
4. F (p.216)
5. F (p.216)
6. T (p.216)
7. T (p.217)
8. F (p.219)
9. T (p.219)
10. T (p.220)
11. F (p.222)
12. F (pp.223-224)
13. T (p.226)
14. T (p.229)
15. F (p.232)
16. T (p.232)
17. F (p.233)
18. T (p.235)
19. T (p.237)
20. F (p.240)

Short Answer Questions:

1. See p.212
2. See pp.216-217
3. See p.232
4. See pp.237-239

Chapter 10

CHAPTER TEN

GLOBAL STRATIFICATION

Multiple Choice:

1. According to the text, which of the following factors reveal(s) the consequences of a global system of inequality?

    A. life expectancy
    B. infant mortality
    C. access to health service
    D. all the above

2. Students at the University of _____ organized a "Why shop?" week in November 1999, posting a Web page with information about college logos, sweatshops, and garment workers in the United States and abroad.

    A. Michigan
    B. Chicago
    C. Florida
    D. Colorado

3. The government estimates _____ percent of U.S. garment shops violate safety and health laws.

    A. 50
    B. 75
    C. 25
    D. 100

4. In China, workers molding Barbie dolls earn _____ per hour, and human rights organizations say violations of basic rights are flagrant.

    A. 50 cents
    B. 75 cents
    C. 25 cents
    D. $1

5. According to the text, in _____ more toys are produced than in any other part of the world.

    A. China
    B. Mexico
    C. Indonesia
    D. Vietnam

6. Approximately how many workers work in toy factories in the United States today?

   A. 13,000
   B. 56,000
   C. 27,000
   D. 92,000

7. In 2002, the per capita GNI of the United States was:

   A. $26,400
   B. $35,400
   C. $42,400
   D. $53,400

8. According to the text, the _____ provides a measure of the relative affluence of those living in the United States.

   A. GNI
   B. GDP
   C. GDI
   D. none of the above

9. According to the text, the United States ranks _____ among the world's wealthiest nations.

   A. tenth
   B. fourth
   C. third
   D. sixth

10. In general, poor, underdeveloped, largely rural countries that have autocratic dictatorships and economies based on farming are considered:

    A. second world countries
    B. third world countries
    C. first world countries
    D. core countries

11. According to the text, the oil-rich countries of the Middle East (i.e., Saudi Arabia) would be categorized as:

    A. first world countries
    B. second world countries
    C. third world countries
    D. none of the above

Chapter 10

12. Besides the United States, most wealthy countries are in:

    A. Eastern Europe
    B. Western Europe
    C. East Asia
    D. Latin America

13. Which of the following countries is(are) considered a semiperipheral country?

    A. Spain
    B. Turkey
    C. Mexico
    D. all of the above

14. Using power as a dimension, the countries of the world can be stratified on three levels. The semi-industrialized countries that represent a kind of middle class in this stratification system are the _____ countries.

    A. "core"
    B. "peripheral"
    C. "semiperipheral"
    D. "hybrid"

15. Which of the following countries is not considered a core country?

    A. Brazil
    B. Great Britain
    C. Japan
    D. Australia

16. The second-world countries are socialist countries without democratically elected governments, including the former Soviet Union, China, Cuba, and:

    A. New Zealand
    B. Venezuela
    C. North Korea
    D. none of the above

17. Which country's elite declared their country a racial democracy from the early stages of national development?

    A. the United States
    B. South Africa
    C. Greece
    D. Brazil

18. Anthony Marx's study of racism around the world has revealed that the creation of arbitrary labels in the long run may lead to:

    A. more racial equality by providing an identity around which political mobilization can take place
    B. substantial guilt by the dominant group once they realize the consequences of labeling
    C. an exodus by members of the minority group to a location where they will be more accepted
    D. group self-hatred by members of the target group toward their own distinctive characteristics

19. _____ in his book titled *How Europe Underdeveloped Africa*, argues that Europeans and North Americans have underdeveloped African nations by exploiting their resources and retarding their economic growth.

    A. Thomas Friedman
    B. Walter Rodney
    C. Anthony Marx
    D. Immanuel Wallerstein

20. According to modernization theory, the economic development of countries stems from _____ change.

    A. technological
    B. political
    C. agricultural
    D. none of the above

21. Modernization theory is similar to the argument of _____, which sees people as poor because they have poor work habits, engage in poor time management, are not willing to defer gratification, and do not save or take advantage of educational opportunities.

    A. dependency
    B. the cycle of poverty
    C. the culture of poverty
    D. none of the above

22. _____ can spur or hinder economic development, especially as they work with private companies, to enact export strategies, restrict imports, or place embargos on the products of nonfavored nations.

    A. NGOs
    B. Governments
    C. Multinational corporations
    D. Transnational corporations

Chapter 10

23. According to _____, the development schemes of the richest countries have resulted in the underdevelopment and poverty of the poor nations.

   A. modernization theory
   B. social exchange theory
   C. dependency theory
   D. functionalism

24. _____, derived from the work of Karl Marx, focuses on explaining the persistence of poverty in the low-income countries as a direct result of their political and economic relationship with wealthy countries.

   A. Dependency theory
   B. Modernization theory
   C. Technological-imperative theory
   D. none of the above

25. According to the text, India was a British colony from 1757 to _____, and during that time, Britain bought cheap _____ from India, made it into cloth in British Mills, and then sold the cloth back to India, making large profits.

   A. 1912; hemp
   B. 1947; hemp
   C. 1912; cotton
   D. 1947; cotton

26. According to the text, which of the following countries was never a colony?

   A. Mexico
   B. Singapore
   C. Ethiopia
   D. Hong Kong

27. Modernization theory examines the factors:

   A. external to an individual country
   B. internal to an individual country
   C. that impact the relationship between countries
   D. none of the above

28. _____ theory was originally developed to explain the historical evolution of global capitalism.

   A. Modernization
   B. Dependency
   C. World systems
   D. Labeling

29. _____ studied two communities of Filipina women to learn how their experiences were part of the system of global stratification.

   A. John Kenneth Galbraith
   B. W. W. Rostow
   C. Paul Samuelson
   D. Rhacel Salazar Parreñas

30. According to the text, the richest countries of the world have _____ percent of the world's total population.

   A. 10
   B. 20
   C. 15
   D. 30

31. _____ theory finds that peripheral countries often benefit by housing low-wage factories and that the core countries are sometimes hurt when jobs move overseas.

   A. Commodity chain
   B. Modernization
   C. Assembly-line
   D. Division of process

32. Three out of _____ people in the world live on less than one dollar per day.

   A. ten
   B. seven
   C. four
   D. none of the above

33. The poorest countries have the _____ birthrates and the _____ death rates.

   A. highest; highest
   B. highest; lowest
   C. lowest; highest
   D. lowest; lowest

34. The populations in the poorest countries live in mostly _____ areas, yet the richest countries are largely _____.

   A. suburban; spread evenly between rural and urban areas
   B. urban; rural
   C. rural; urbanized
   D. none of the above

Chapter 10

35. According to the text, life expectancy in Afghanistan is approximately _____ years.

    A. 69
    B. 73
    C. 77
    D. 43

36. According to the text, _____ percent of women are enrolled in elementary school in Afghanistan.

    A. 95
    B. 15
    C. 74
    D. 100

37. According to the text, _____ percent of the population in Mexico have access to safe water.

    A. 15
    B. 74
    C. 13
    D. 86

38. Which of the following countries has the third largest population in the world?

    A. Brazil
    B. China
    C. India
    D. the United States

39. The emission of carbon dioxide (i.e., the overproduction of "greenhouse gas") from the burning of fossil fuels is most severe in countries that:

    A. are far from the equator
    B. are close to the equator
    C. use the least energy
    D. use the most energy

40. Although high-income countries have only _____ percent of the world population, together they use more than half of the world's energy.

    A. 30
    B. 20
    C. 5
    D. 15

41. According to the text, in every nation, the _____ index is less than the general human development index.

   A. gender equality index
   B. gender development index
   C. gender gap index
   D. gender disparity index

42. According to the text, a study of al Qaeda terrorists finds that the leaders tend to come from _____-class backgrounds.

   A. lower
   B. upper
   C. middle
   D. none of the above

43. _____ poverty is a situation in which individuals live on less than $365 a year, meaning that people at this level of poverty live on approximately $1 a day.

   A. Absolute
   B. Extreme
   C. Relative
   D. none of the above

44. The United Nations Commission on the Status of Women, women constitute almost _____ percent of the world's population.

   A. 45
   B. 50
   C. 60
   D. 55

45. According to the text, the _____ fertility rates are found in the areas with the _____ poverty.

   A. lowest; greatest
   B. highest; greatest
   C. highest; least
   D. lowest; least

46. The United Nations estimates that there are _____ children between the ages of five and fourteen in the paid labor force throughout the world.

   A. 127 million
   B. 211 million
   C. 13 million
   D. 46 million

Chapter 10

47. What set off a collapse of national economies that created massive amounts of poverty and starvation?

    A. worldwide compression of economic activity
    B. new weather patterns
    C. lack of food crops
    D. high fertility rates

48. Countries, such as Korea, Malaysia, Thailand, Taiwan, and Singapore, that have shown rapid growth and have emerged as developed countries are:

    A. Newly Developed Countries (NDC)
    B. Economically Revised Nations (ERN)
    C. Newly Industrializing Countries (NIC)
    D. Urbanizing, Industrializing Nations (UIN)

49. According to the text, the new economic push globally is to:

    A. further develop the world as one large capital market
    B. eliminate hunger in third-world countries
    C. eliminate poverty throughout the world by redistributing profits of the world system to have-not countries
    D. provide education, hospitals, roads, and other social services to upgrade the overall quality of life in undeveloped and underdeveloped countries

50. Which of the following statements represents a negative aspect of building capital markets?

    A. the development of new markets
    B. capital markets also create poverty
    C. the ability to harness new resources
    D. all the above

True/False Questions:

T F 1. Looking at the world today reminds us that nations are independent of economic and sociological processes.

T F 2. Per capita GNI is reliable only in countries that are based on a cash economy.

T F 3. According to information presented in the text, Luxembourg is the poorest country with the lowest GNI per capita.

T F 4. Based on reliable economic data, most of the world's poorest countries are in South America.

T F 5. At the top of the world stratification system are the peripheral countries.

T F 6. Dependency theory attempts to explain the persistence of poverty in the low-income countries as a direct result of their political and economic relationship on the wealthy countries.

T F 7. According to dependency theory, multinational corporations assist poor nations in becoming economically self-sustaining.

T F 8. World systems theory was originally developed to explain the historical evolution of global capitalism, but now also explains how differential profits are attached to the production of goods and services in the world market.

T F 9. The richest countries of the world make up only 15 percent of the world's population.

T F 10. Rapid population growth as a result of high fertility rates makes a large difference in the quality of life of the country.

T F 11. Educational attainment in third world countries is along the same level as other countries.

T F 12. The gender development index is calculated based on gender inequalities in life expectancy, educational attainment, and income for different countries.

T F 13. Currently, the relative poverty rate in the United States for a family of four is $18,307 (2002).

T F 14. According to the text, the human poverty index, is a multidimensional measure of poverty.

T F 15. There is no connection between high fertility rates and the degree of women's empowerment in society.

T F 16. Latin America has an estimated 13 million street children, some as young as six years old.

T F 17. Due to the growth of food sources worldwide, children throughout the world do not suffer from malnutrition.

T F 18. Poverty is not caused because people are lazy or uninterested in working.

T F 19. In Latin America, the poor have flooded to the cities, hoping to find work, whereas in Africa, they did the opposite.

T F 20. Market economies do not create opportunities to become wealthy, for individuals or nations.

Chapter 10

Short Answer Questions:

1. Compare and contrast between core, semiperipheral, and peripheral countries.
2. Compare and contrast between modernization theory and dependency theory.
3. Compare and contrast between absolute and extreme poverty.
4. Discuss the causes of world poverty presented in the text.

Global Stratification

ANSWER FOR CHAPTER TEN

Multiple Choice Questions:

1. D (p.246)
2. D (p.247)
3. B (p.247)
4. C (p.247)
5. A (p.247)
6. C (p.248)
7. B (p.249)
8. A (p.249)
9. D (p.249)
10. B (p.250)
11. D (p.250)
12. B (p.250)
13. D (p.250)
14. C (p.250)
15. A (p.250)
16. C (p.250)
17. D (p.251)
18. A (p.251)
19. B (p.251)
20. A (p.252)
21. C (p.252)
22. B (p.253)
23. C (p.253)
24. A (p.253)
25. D (p.253)
26. C (p.254)
27. B (p.254)
28. C (p.254)
29. D (p.255)
30. C (p.256)
31. A (p.256)
32. B (p.256)
33. A (p.256)
34. C (p.257)
35. D (p.257, Table 10.2)
36. B (p.257, Table 10.2)
37. D (p.257, Table 10.2)
38. D (p.257)
39. D (p.258)
40. D (p.258)
41. B (p.259)
42. C (p.260)
43. A (p.260)
44. C (p.262)
45. B (p.263)
46. B (p.264)
47. A (p.265)
48. C (p.265)
49. A (p.265)
50. B (p.265)

True/False Questions:

1. F (p.245)
2. T (p.248)
3. F (p.249)
4. F (p.249)
5. F (p.250)
6. T (p.253)
7. F (p.253)
8. T (p.254)
9. T (p.256)
10. T (p.257)
11. F (p.258)
12. T (p.258)
13. T (p.260)
14. T (p.261)
15. F (p.263)
16. T (p.264)
17. F (p.264)
18. T (p.264)
19. T (p.265)
20. F (p.265)

Short Answer Questions:

1. See p.250
2. See pp.252-254
3. See p.260
4. See p.264-265

Chapter 11

CHAPTER ELEVEN

RACE AND ETHNICITY

Multiple Choice:

1. A social category of people who share a common culture, for example, a common language or dialect, a common religion, and common norms, practices, and customs is:

    A. a race
    B. an assimilated minority group
    C. a culturally amalgamated group
    D. an ethnic group

2. More than anything else, what made Italian Americans an ethnic group is their:

    A. physical similarity
    B. commonly felt social and cultural bonds, arising from historical experiences
    C. residential isolation from other groups
    D. all of the above were of equal importance

3. From a sociological perspective, how racial groups are defined is a _____ process.

    A. genetic
    B. biological
    C. psychological
    D. social

4. "Race" in Brazil is defined by one's:

    A. skin color
    B. facial features
    C. social class
    D. none of the above

5. The vast majority of U.S. Blacks would not be considered Black in:

    A. Canada
    B. Mexico
    C. Great Britain
    D. Brazil

6. The government makes Indian tribes prove themselves as tribes through a complex set of federal regulations called the:

   A. "federal acknowledgment process"
   B. "tribal membership process"
   C. "tribal acknowledgment process"
   D. "federal membership process"

7. According to the text, a century ago the _____ were called "Negroes turned inside out."

   A. British
   B. Scottish
   C. Irish
   D. Swedish

8. The _____ effect refers to a situation where all members of any out-group are perceived by an individual to be similar or even identical to each other, and differences among them are perceived to be minor or nonexistent.

   A. in-group homogeneity
   B. out-group homogeneity
   C. in-group heterogeneity
   D. out-group heterogeneity

9. The process by which a group comes to be defined as a race, through support by official institutions, such as the law and schools, is:

   A. racial formation
   B. aversive racism
   C. racial stratification
   D. dominant racism

10. Any distinct group in society that shares common group characteristics and is forced to occupy low status in society because of prejudice and discrimination is a(n):

    A. ethnic group
    B. minority group
    C. race
    D. stereotyped group

11. The group that assigns a racial or ethnic group to subordinate status in society is a(n):

    A. dominant group
    B. urban underclass
    C. social majority
    D. none of the above

Chapter 11

12. An oversimplified set of beliefs about members of a social group or social stratum that is used to categorize individuals of that group is a(n):

   A. prejudice
   B. stereotype
   C. discrimination
   D. affirmative action

13. Which of the following groups escaped the process of categorization and stereotyping in the United States?

   A. Germans
   B. Swedes
   C. Polish
   D. no group escaped this process

14. According to the text, _____-class people are more likely to attribute their status to an external societal factor.

   A. middle
   B. upper
   C. lower
   D. none of the above

15. According to the text, _____-class people are more likely to attribute the low status of a lower-class person to something internal, such as lack of willpower.

   A. middle
   B. upper
   C. lower
   D. none of the above

16. Less prejudiced Whites have very different causes they associate with urban poverty among Blacks than more racially prejudiced White persons. Less prejudiced Whites are least apt to blame _____ for urban poverty among Blacks.

   A. social structure
   B. racism
   C. "laziness"
   D. discrimination

17. _____ is the evaluation of a social group, and the individuals within that group, based on conceptions about the social group that are held together despite facts that contradict it, and that involves both a prejudgment and misjudgment.

   A. Discrimination
   B. Prejudice
   C. Racism
   D. Ethnocentrism

18. Prejudice is frequently reflected in ethnocentrism. Generally, ethnocentric persons feel that:

    A. their group is superior to all other groups
    B. a cultural trait must be judged based on the context in which it appears
    C. assimilation of minorities strengthens one's society by promoting a sense of social solidarity
    D. all of the above are true

19. In the United States, poverty is highest among:

    A. Japanese Americans
    B. White Americans
    C. Hispanic Americans
    D. African Americans

20. Racism includes:

    A. attitudes
    B. behaviors
    C. perception
    D. all of the above

21. Overt negative and unequal treatment of the members of some social group or stratum solely because of their membership in that group or stratum is:

    A. prejudice
    B. racism
    C. discrimination
    D. ethnocentrism

22. An illegal practice in which an entire minority neighborhood is designated "no-loan" is called:

    A. loanmandering
    B. gerrymandering
    C. red lining
    D. segregation lining

23. The spatial segregation of racial and ethnic groups into different housing areas of the United States is:

    A. de jure segregation
    B. de facto segregation
    C. residential segregation
    D. all of the above

24. Currently, the rate of segregation of Blacks and Hispanics in U.S. cities is:

    A. increasing
    B. decreasing
    C. the same as it was decades ago
    D. none of the above

Chapter 11

25. A close relative of laissez-faire racism is:

   A. aversive racism
   B. color-blind racism
   C. old-fashioned racism
   D. none of the above

26. Obvious, overt racism is called _____ racism.

   A. traditional
   B. aversive
   C. benign
   D. de facto

27. *Symbolic racism* is another name for:

   A. institutional racism
   B. old-fashioned racism
   C. aversive racism
   D. laissez-faire racism

28. Racial profiling is an example of:

   A. aversive racism
   B. old-fashioned racism
   C. institutional racism
   D. laissez-faire racism

29. The Fair Housing Act of _____, intended to decrease housing discrimination.

   A. 1994
   B. 1954
   C. 1968
   D. 1972

30. The theory that argues that historically members of the dominant group in the United States have harbored various frustrations in their desire to achieve social and economic success, which they direct toward minority group members, is the:

   A. frustration-compensation complex
   B. scapegoat theory
   C. contact theory
   D. conflict theory

31. Which of the following theories addresses the role of social interaction in reducing racial and ethnic hostility?

   A. social exchange theory
   B. functionalism
   C. conflict theory
   D. symbolic interaction theory

112

Race and Ethnicity

32. Which of the following theorists is most likely to argue that class inequality must be reduced to lessen racial and ethnic conflict in society?

    A. conflict theory
    B. functionalism
    C. social exchange theory
    D. symbolic interaction theory

33. Sociologist William Julius Wilson argues that:

    A. being disadvantaged in the U.S. is more a matter of race than class
    B. directly addressing the question of race forthrightly is the only way to solve the country's racial problems
    C. "gendered racism" is a myth
    D. class and changes in the economic structure are more important than race per se in shaping the life changes of different groups, although they both are important

34. According to the text, the Sioux reservation was established in:

    A. 1834
    B. 1889
    C. 1800
    D. 1792

35. Today, about _____ percent of all Native Americans live on or near a reservation.

    A. 55
    B. 75
    C. 45
    D. 65

36. The slave system also involved the domination of men over:

    A. machine
    B. status
    C. women
    D. none of the above

37. The island of _____ was ceded to the United States by Spain in 1899.

    A. Guam
    B. Puerto Rico
    C. Cayman Islands
    D. none of the above

38. In _____, the Jones Act extended U.S. citizenship to Puerto Ricans.

    A. 1952
    B. 1948
    C. 1899
    D. 1917

Chapter 11

39. In _____, the United States established the Commonwealth of Puerto Rico, with its own constitution.

   A. 1917
   B. 1948
   C. 1899
   D. 1952

40. The U.S. government encouraged the sterilization of Puerto Rican women, and by 1974, more than _____ percent of the women of reproductive age in Puerto Rico had been sterilized.

   A. 8
   B. 21
   C. 12
   D. 37

41. In the class system that emerged within the Chicano community during the Golden Age of the Ranchos, the upper strata was comprised of:

   A. elite ranchers
   B. mission farmers
   C. government administrators
   D. all of the above

42. The most recent wave of Cuban immigration came in 1980, when the Cuban government opened the port of:

   A. Talcahuano
   B. Mariel
   C. Rio Piedras
   D. Carolina

43. In _____, the U.S. federal government passed the Chinese Exclusion Act, which banned further immigration of unskilled Chinese laborers.

   A. 1890
   B. 1917
   C. 1924
   D. 1882

44. The second generation of Japanese Americans is called:

   A. *Nee*
   B. *Issei*
   C. *Sansei*
   D. *Nisei*

Race and Ethnicity

45. By executive order of President _____, much of the West Coast Japanese American population.

    A. John F. Kennedy
    B. Franklin D. Roosevelt
    C. Lyndon B. Johnson
    D. Harry Truman

46. In _____, the U.S. Supreme Court allowed Japanese Americans the right to file suit for monetary reparations.

    A. 1973
    B. 1952
    C. 1986
    D. 1968

47. More than _____ percent of the world's Jewish population lives in the United States.

    A. 50
    B. 40
    C. 30
    D. 20

48. If one is both a woman and a minority, one is often confronted by a form of discrimination known as:

    A. multiple trait effect
    B. double discrimination effect
    C. double jeopardy effect
    D. triple jeopardy effect

49. The major force behind the most progressive social change in race relations in the U.S. was:

    A. an upturn is the economy during the 1960's and 1970's
    B. the end of the "Cold War"
    C. the civil rights movement
    D. the "Brown decision" ruling by the U.S. Supreme Court

50. The Black power movement of the late 1960's advocated fighting oppression with:

    A. armed revolution
    B. a philosophy that focused on increased educational opportunity for Blacks
    C. legal action that would affect political reform
    D. peaceful protest based on principles espoused by Black Muslim leadership

Chapter 11

True/False Questions:

T   F   1.   Within sociology, the terms *ethnic, race, minority,* and *dominant group* have the same meaning.

T   F   2.   Racialization is a process whereby some social categories such as social class or nationality takes on what are perceived in the society to be race characteristics.

T   F   3.   Race has no distinction in biological and sociological importance in society.

T   F   4.   Racial formation is the composite of a particular ethnic group.

T   F   5.   As a general rule in society at large, people do not categorize others.

T   F   6.   The salience principle states that we categorize people on the basis of what appears initially prominent and obvious.

T   F   7.   Prejudice based on race or ethnicity is called racial-ethnic prejudice.

T   F   8.   Ethnocentrism is a belief that allows for the acceptance of others no matter what their ethnicity, race, age, gender, and religion.

T   F   9.   Racial ethnic discrimination is unequal treatment of a person on the basis of race or ethnicity.

T   F   10.   Housing discrimination is illegal under U.S. law.

T   F   11.   Overt racism is called aversive racism.

T   F   12.   Racism that is subtle, covert, and not obvious is termed dominative racism.

T   F   13.   Racism is reflected not only in individual overt behavior, but also in society's institutions, called institutional racism.

T   F   14.   Assimilation is a process by which a minority becomes socially, economically, and culturally absorbed within the dominant society.

T   F   15.   Conflict theory is the basis for understanding race and ethnicity as socially constructed categories.

T   F   16.   The population of Latin Americans has grown considerably over the past few decades, with the largest increase among Mexican Americans.

T   F   17.   The original WASP immigrants were skilled workers imbued with the work ethic of Karl Marx.

T   F   18.   The assimilationist believes that to overcome adversity and oppression, the minority person needs to adapt to the dominant society.

Race and Ethnicity

T   F   19.   The effects of race and class can interact with the effect of gender, yielding a triple jeopardy effect.

T   F   20.   Affirmative action has been widely accepted throughout U.S. society.

Short Answer Questions:

1. Define racialization and provide examples.
2. Compare and contrast between aversive and laissez-faire racism.
3. Compare and contrast between assimilation and pluralism.
4. Compare and contrast between color-blind and race-specific policies.

Chapter 11

ANSWERS FOR CHAPTER ELEVEN

Multiple Choice Questions:

1. D (p.270)
2. B (p.270)
3. D (p.271)
4. C (p.272)
5. D (p.272)
6. A (p.274)
7. C (p.274)
8. B (p.274)
9. A (p.274)
10. B (p.274)
11. D (p.275)
12. B (p.275)
13. D (p.275)
14. C (p.276)
15. A (p.276)
16. C (p.276)
17. B (p.276)
18. A (p.277)
19. D (p.278)
20. D (p.278)
21. C (p.278)
22. C (p.278)
23. C (p.278)
24. A (p.278)
25. B (p.279)
26. A (p.279)
27. D (p.279)
28. C (p.280)
29. C (p.280)
30. B (p.281)
31. D (p.281)
32. A (p.282)
33. D (p.282)
34. B (p.284)
35. A (p.284)
36. C (p.285)
37. B (p.286)
38. D (p.286)
39. D (p.286)
40. D (p.286)
41. D (p.286)
42. B (p.287)
43. D (p.287)
44. D (p.287)
45. B (p.287)
46. C (p.288)
47. B (p.290)
48. C (p.292)
49. C (p.293)
50. A (pp.294-295)

True/False Questions:

1. F (p.270)
2. T (p.272)
3. F (pp.272-274)
4. F (p.274)
5. F (p.274)
6. T (p.275)
7. T (p.277)
8. F (p.277)
9. T (p.278)
10. T (p.278)
11. F (p.279)
12. F (p.279)
13. T (p.279)
14. T (p.281)
15. F (p.282)
16. T (p.286)
17. F (p.289)
18. T (p.291)
19. T (p.292)
20. F (pp.295-296)

Short Answer Questions:

1. See p.272
2. See p.279
3. See p.291
4. See p.295

CHAPTER TWELVE

GENDER

Multiple Choice:

1. Long after passage of the Equal Pay Act in 1963, men still earn _____ percent more than women, counting only those working year-round and full-time.

    A. 12
    B. 45
    C. 36
    D. 23

2. Which of the following concepts is more significant for sociologists?

    A. female
    B. male
    C. sex
    D. gender

3. The *Berdaches* in _____ society were anatomically normal men who were defined as a third gender.

    A. Sioux
    B. Navajo
    C. Seminoles
    D. Maya

4. Two X chromosomes make a:

    A. male
    B. hermaphrodite
    C. female
    D. none of the above

5. An X and a Y chromosome makes a:

    A. male
    B. hermaphrodite
    C. female
    D. none of the above

Chapter 12

6. According to the text, gender identity is formed through:

   A. social interaction
   B. biological processes
   C. the deconstruction of personality structures
   D. none of the above

7. Studies consistently show that _____ interrupt in conversation – especially when talking to _____.

   A. women; men
   B. men; men
   C. men; women
   D. women; women

8. Gender socialization is so effective that, as early as _____ months of age, toddlers have learned to play with presumed gender-appropriate toys.

   A. twelve
   B. eighteen
   C. fifteen
   D. nine

9. Mexican married couples who migrate to the United States tend to adopt more _____ family roles.

   A. egalitarian
   B. patriarchal
   C. matriarchal
   D. traditional

10. In general, _____ in school get more attention, even if it is _____ attention.

    A. girls; negative
    B. boys; negative
    C. girls; positive
    D. boys; positive

11. The major Judeo-Christian religions in the United States place strong emphasis on gender differences, with explicit affirmation of the:

    A. equal distribution of authority
    B. authority of women over men
    C. authority of men over women
    D. none of the above

12. According to the text, Harlequin novels account for _____ percent of all paperback sales.

    A. 60
    B. 20
    C. 50
    D. 30

13. Harlequin, only one of many romance publishers, is estimated to have more than _____ loyal readers.

    A. 52 million
    B. 188 million
    C. 14 million
    D. 87 million

14. Research on the racial expectations of different groups indicates that:

    A. Hispanic men and women are more likely than White men and women to select different characteristics in men and women as desirable
    B. African American men are more likely than African American women to support feminism and egalitarian views of men and women's roles
    C. comparing Hispanics, Whites, and African Americans, it is African Americans who are most likely to find value in both sexes being assertive, athletic, and self-reliant
    D. all of the above are true

15. According to the text, in the U.S. working-class women are more likely to be seen as:

    A. "frigid"
    B. "slutty"
    C. "cold"
    D. "spoiled"

16. Which of the following groups is most likely to be stereotyped as "hot"?

    A. African American women
    B. Jewish American women
    C. Asian American women
    D. Latinas

17. Which of the following groups of women is most likely to be also socialized to be self-sufficient and independent?

    A. White women
    B. Asian American women
    C. African American women
    D. Latinas

Chapter 12

18. According to the text, the definition of manhood among _____ is more multidimensional than cultural stereotypes suggest.

   A. Latinos
   B. Japanese American men
   C. Chinese American men
   D. Jewish American men

19. Homophobia plays an important role in gender socialization because it:

   A. affirms the male's role in society
   B. helps to ostracize homosexuals from the dominant society
   C. encourages stricter conformity to traditional gender expectations
   D. none of the above

20. Homophobia is a _____ attitude.

   A. natural
   B. instinctive
   C. innate
   D. learned

21. A good example of a gendered institution is:

   A. hospitals
   B. elementary schools
   C. all-male military academies
   D. realty agencies

22. According to the text, _____ of the illiterate people in the world are women.

   A. two-thirds
   B. one quarter
   C. half
   D. one fifth

23. In which of the following countries was "gender apartheid" most noteworthy during the last decade?

   A. Sweden
   B. Afghanistan
   C. Iceland
   D. Bulgaria

24. Which of the following is an example of an "ideology"?

   A. sexism
   B. racism
   C. ageism
   D. all of the above

25. According to the text, in which country is the wage gap between men and women the smallest?

   A. Austria
   B. Turkey
   C. Denmark
   D. the United States

26. According to the text, in which country is the wage gap between men and women the largest?

   A. Bangladesh
   B. Turkey
   C. Denmark
   D. the United States

27. Gender stratification can be extreme, such as was the case when the Taliban seized power in:

   A. Afghanistan
   B. Bangladesh
   C. Iraq
   D. Kuwait

28. Women with college degrees earn the equivalent of men who have only:

   A. an 8th grade education
   B. some high school
   C. high school diploma
   D. some college education

29. Which of the following factors is(are) considered a human capital variable(s)?

   A. education
   B. marital status
   C. age
   D. all of the above

30. The _____ is the first federal law to require that men and women receive equal pay for equal work.

   A. Pay equity Act of 1958
   B. Equal Pay Act of 1954
   C. Pay equity Act of 1966
   D. Equal Pay Act of 1963

31. Which of the following sectors of the labor market has greater wage inequality between men and women?

   A. formal sector
   B. informal sector
   C. business sector
   D. industrial sectors

Chapter 12

32. According to the text, the greater the proportion of _____ in a given occupation, the _____ the pay.

    A. men; lower
    B. women; higher
    C. women; lower
    D. none of the above

33. Studies find that workers in jobs requiring _____ skills have the lowest pay, even when their education and experience are comparable to workers in other jobs.

    A. analytical
    B. physical
    C. nurturing social
    D. none of the above

34. The majority of women work in occupation where most of the other workers are _____, and the majority of men work mostly with _____.

    A. women; men
    B. men; women
    C. their subordinates; superiors
    D. none of the above

35. By current estimates, at least _____ percent of men (or women) would have to change occupations to achieve occupational balance by gender.

    A. 78
    B. 53
    C. 64
    D. 44

36. _____ percent of elementary school teachers are women.

    A. 99
    B. 91
    C. 83
    D. 79

37. Many men are now much more involved in housework and childcare than has been true in the past, although most of this work still falls on women – a phenomenon that has been labeled:

    A. "the second shift"
    B "the 36-hour day"
    C. "the feminine imperative"
    D. "the male prerogative"

38. According to the text, Hispanic women earn only _____ percent of what White men earn.

   A. 64
   B. 84
   C. 52
   D. 83

39. Overall, women's income is _____ percent of the income's of men.

   A. 64
   B. 74
   C. 52
   D. 83

40. According to the text, less than _____ countries have elected a woman president.

   A. 35
   B. 30
   C. 25
   D. 20

41. In May 2000, the U.S. Supreme Court ruled that victims of rape, domestic violence, and other crimes motivated by gender could:

   A. testify against their attackers in civil courts
   B. not testify against their attackers in civil courts
   C. sue their attackers through civil courts
   D. not sue their attackers through civil courts

42. According to _____ feminism, race, class, and gender intersect to form a matrix of domination.

   A. socialist
   B. radical
   C. liberal
   D. multiracial

43. According to _____ feminism, gender inequality stems from class relations.

   A. radical
   B. socialist
   C. liberal
   D. multiracial

Chapter 12

44. According to _____ feminism, liberation comes as women organize on their own behalf.

    A. socialist
    B. multiracial
    C. liberal
    D. radical

45. According to _____ feminism, change is accomplished through legal reform and attitudinal change.

    A. socialist
    B. radical
    C. multiracial
    D. liberal

46. Feminist sociologists have been especially influenced by:

    A. functionalism
    B. conflict theory
    C. social exchange theory
    D. none of the above

47. About _____ percent of men now disapprove of women being employed.

    A. 10
    B. 20
    C. 30
    D. 40

48. _____ percent of women say that making laws to establish equal pay should be a legislative priority.

    A. Sixty-seven
    B. Seventy-seven
    C. Eighty-seven
    D. Ninety-seven

49. The passage of the Civil Rights Act and Title VII opened up new opportunities to women in:

    A. employment
    B. education
    C. both A and B are true
    D. none of the above

Gender

50. In _____, the U.S. Supreme Court further strengthened Title IX by refusing to rule on a case that supported the principles embedded in Title IX.

    A. 1997
    B. 1994
    C. 1989
    D. 2001

True/False Questions:

T F 1. Two X chromosomes make a male.

T F 2. Hermaphrodites are also called intersex persons.

T F 3. Arguments based on biological determinism assume that differences between women and men are learned.

T F 4. Transgendered people are those who are biological deviates.

T F 5. Gender identity is a gender socialization process to help homophobics.

T F 6. Agents of gender socialization comprise the family, schools, religion, and the mass media.

T F 7. Conforming to gender expectations can have negative consequences for women.

T F 8. African American men are less likely than White men to emphasize their importance as the breadwinner.

T F 9. Gender stratification refers to the hierarchical distribution of social and economic resources according to gender.

T F 10. Sexism is an ideology that suggests women are controlled because of the significance given to differences between the sexes.

T F 11. Women with college degrees earn the equivalent of men who have college degrees.

T F 12. Matriarchy exists, but not as a mirror image of patriarchy.

T F 13. The discrimination explanation of the gender wage gap argues that dominant groups will use their position of power to perpetuate their advantage.

T F 14. For all women, perceptions of gender appropriate behavior influence the likelihood of success at work.

T F 15. The glass ceiling exists at every level where men and women work together.

Chapter 12

T F 16. To help working women China has developed extensive child care facilities and a paid maternity leave.

T F 17. Feminist scholars strongly oppose the "doing gender" theoretical perspective on gender.

T F 18. Feminist theory refers to analyses that seek to understand the position of women in society for the purposes of bringing about liberating changes.

T F 19. Radical feminism interprets patriarchy as the primary cause of women's oppression.

T F 20. The implementation of Title IX legislation has established equity in women's sports at the college level.

Short Answer Questions:

1. Compare and contrast between sex and gender.
2. Define homophobia. Provide examples.
3. Compare and contrast between patriarchy and matriarchy. Provide examples.
4. Compare and contrast between liberal and radical feminism.

ANSWERS FOR CHAPTER TWELVE

Multiple Choice Questions:

1. D (p.301)
2. D (p.302)
3. B (p.302)
4. C (p.303)
5. A (p.303)
6. A (p.305)
7. C (p.305)
8. B (p.305)
9. A (p.306)
10. B (p.307)
11. C (p.308)
12. D (p.310)
13. C (p.310)
14. C (p.312)
15. B (p.312)
16. D (p.312)
17. C (p.312)
18. A (p.313)
19. C (p.313)
20. D (p.314)
21. C (p.314)
22. A (p.315)
23. B (p.316)
24. D (p.316)
25. B (p.316, Figure 12.4)
26. A (p.316, Figure 12.4)
27. A (p.316)
28. D (p.317)
29. D (p.318)
30. D (p.318)
31. B (p.319)
32. C (p.319)
33. C (p.319)
34. A (p.320)
35. B (p.320)
36. C (pp.320-321)
37. A (p.322)
38. C (p.323)
39. B (p.323)
40. D (p.324)
41. D (p.325)
42. D (p.326, Table 12.2)
43. B (p.326, Table 12.2)
44. D (p.326, Table 12.2)
45. D (p.326, Table 12.2)
46. B (p.326)
47. B (p.328)
48. C (p.328)
49. D (p.328)
50. A (p.329)

True/False Questions:

1. F (p.303)
2. T (p.303)
3. F (p.304)
4. F (p.304)
5. F (p.305)
6. T (p.305)
7. T (p.310)
8. F (p.313)
9. T (p.315)
10. T (p.316)
11. F (p.317)
12. T (p.317)
13. T (p.318)
14. T (p.321)
15. T (p.322)
16. T (p.323)
17. F (p.325)
18. T (p.327)
19. T (p.327)
20. F (p.328)

Short Answer Questions:

1. See p.302
2. See pp.313-314
3. See p.317
4. See p.327

Chapter 13

CHAPTER THIRTEEN

SEXUALITY

Multiple Choice:

1. Gays and lesbians won a major civil rights victory in 1996 when the Supreme Court ruled, in _____, that gays and lesbians cannot be denied equal protection under the law.

    A. *Eisenstadt v. Baird*
    B. *Griswold v. Connecticut*
    C. *Brown v. Board of Education of Topeka*
    D. *Romer v. Evans*

2. Despite the serious scientific flaws in studies supporting an alleged biological basis to homosexuality, _____ of the public believes that homosexuality is a biological trait.

    A. half
    B. one-third
    C. one-quarter
    D. two-thirds

3. The _____ of India are a religious community of men who are born male but come to think of themselves as neither men nor women.

    A. *Berdaches*
    B. *Xaniths*
    C. *hijras*
    D. *Sansei*

4. Whereas Western culture emphasizes dichotomous and separate identities for women and men, _____ religion values the ambiguity of in-between sexual categories, holding that all persons have both male and female principles.

    A. Shinto
    B. Muslim
    C. Buddhist
    D. Hindu

5. Which of the following statements about the social and cultural basis of sexuality is (are) true?

    A. sexual identity is learned
    B. human sexual attitudes vary in different cultural contexts
    C. sexual attitudes and behavior change over time
    D. all of the above

Sexuality

6. The scientific study of sex and sexuality is:

   A. eugenics
   B. sexual politics
   C. sexology
   D. biology

7. In the 1940s it was _____ who first reported that 33 percent of women and 71 percent of men engaged in premarital intercourse despite public belief to the contrary.

   A. Havelock Ellis
   B. Alfred Kinsey
   C. Sigmund Freud
   D. William Masters & Virginia Johnson

8. According to the text, surveys now show that at least _____ of men report having sex before marriage.

   A. 80
   B. 70
   C. 60
   D. 50

9. The theorist who presented a development model of sexuality, and argued that sexual expression began in childhood and developed over the course of the life cycle through several stages of psychosexual development is:

   A. Havelock Ellis
   B. Sigmund Freud
   C. Alfred Kinsey
   D. Janet Lee

10. Kinsey reported that _____ percent of men had experienced homosexual contact resulting in orgasm at some point in their lives.

    A. 43
    B. 25
    C. 12
    D. 37

11. The most extensive and thorough study of human sexual response ever done was completed by:

    A. Havelock Ellis
    B. Janet Lee
    C. Masters and Johnson
    D. Simon LeVay

Chapter 13

12. According to Laumann and colleagues, about _____ percent of adults report having no sexual partners in the past year.

   A. 15
   B. 22
   C. 9
   D. 36

13. The most significant predictor(s) of the frequency of sex is(are):

   A. race
   B. religion
   C. age and marital status
   D. educational level

14. According to _____ theory, sexual identity is socially constructed.

   A. queer theory
   B. conflict
   C. functionalist
   D. symbolic interaction

15. According to _____ theory, multiple forms of sexual identity are possible.

   A. queer
   B. symbolic interaction
   C. conflict
   D. functionalist

16. According to _____ theory, deviant sexual identities contribute to social disorder.

   A. queer
   B. functionalist
   C. symbolic interaction
   D. conflict

17. According to _____ theory, sexual norms are frequently contested by those who are subordinated by dominant groups.

   A. conflict
   B. functionalist
   C. symbolic interaction
   D. queer

18. Sexual _____ implies that sexual _____ is a mere choice.

   A. course; inclination
   B. orientation; preference
   C. preference; orientation
   D. none of the above

Sexuality

19. "Coming out" is:

    A. the process of defining oneself as gay or lesbian
    B. a series of events and redefinitions in which a person comes to see himself or herself as having a gay identity
    C. often a difficult process for the individual, both psychologically and socially
    D. all of the above

20. Sexual politics refers to:

    A. power within individual relationships
    B. both power in individual relationships and the link that exists between sexual exploitation and the distribution of power in society
    C. how sexual exploitation of women is linked to the distribution of power in society
    D. the statistical analysis of elections involving women and men at the federal, state, and local level

21. _____ theory challenges the "either/or" thinking that one is either gay or straight.

    A. Queer
    B. Functionalist
    C. Radical
    D. Social exchange

22. The "double standard" is the idea that for sexual behavior:

    A. different standards apply to men than apply to women
    B. the same standards apply to men and women
    C. both mothers and fathers influence our beliefs
    D. women should have more partners than men

23. The fear and hatred of homosexuality is called:

    A. compulsory heterosexuality
    B. bisexuality
    C. sexology
    D. homophobia

24. _____ is the institution of heterosexuality as the only legitimate sexual orientation, which is reflected in the unequal distribution of privileges to people presumed to be heterosexual.

    A. Compulsory heterosexuality
    B. Heterosexism
    C. Sexual politics
    D. Homophobia

Chapter 13

25. The _____ of Oman, a strictly gender-segregated Islamic society, are men who are homosexual prostitutes.

    A. *Berdaches*
    B. *hijras*
    C. *Xaniths*
    D. *Sansei*

26. According to the text, in the _____ culture, no man is seen as culminating his manhood until he marries and "deflowers" his bride.

    A. Navajo
    B. Hindu
    C. Oman
    D. Sioux

27. The use of women worldwide as sex workers in an institutional context where sex is a commodity used to promote tourism and cater to business and military men is the:

    A. "traffic in women"
    B. international sex trade
    C. capitalist system
    D. both A and B

28. It was not until _____ that the U.S. Supreme Court defined the use of birth control by married people as a right not a crime.

    A. 1972
    B. 1965
    C. 1959
    D. 1945

29. An example of a "personal trouble" that has its origins in the structure of society is:

    A. reproductive technology
    B. pornography
    C. teen pregnancy
    D. all of the above are personal troubles that have their origins in the structure of society

30. According to the text, Margaret Sanger made _____ part of her birth control campaign because she wanted poor women to have fewer children so that their children would not be exploited as workers.

    A. eugenics
    B. capitalism
    C. communism
    D. none of the above

Sexuality

31. Eugenics:

    A. refers to a new reproductive technology which allows previously infertile couples to have children
    B. sought to apply scientific principles of genetic selection to "improve" the offspring of the human race
    C. refers to the scientific study of human sexual behavior
    D. none of the above

32. Data on abortion show that:

    A. women between age 15 and 19 are most likely to get abortions
    B. Catholic women have a much lower rate of abortion than other women
    C. today, the overall abortion rate is the highest it has ever been
    D. The overall abortion rate has increased dramatically since 1980

33. Which of the following statements is (are) true?

    A. Unmarried people have always had the legal right to birth control in the U.S.
    B. The abortion rate has increased steadily since the U.S. Supreme Court first established the right to abortion in 1973
    C. Typically, only the upper class can afford new, expensive reproductive technologies
    D. B and C

34. State policy in _____ encourages families to have only one child.

    A. India
    B. Japan
    C. China
    D. Nigeria

35. According to the text, _____ percent of the U.S. public think abortion should be legal in any or most circumstances.

    A. 58
    B. 41
    C. 22
    D. 17

36. Only _____ percent of the U.S. public think abortion should be illegal in all circumstances.

    A. 58
    B. 22
    C. 17
    D. 41

Chapter 13

37. Public agitation over pornography involves those who think it is solidly protected by the _____ Amendment, those who want it strictly controlled.

   A. First
   B. Eight
   C. Second
   D. Fourth

38. Each year about _____ teenage girls have babies in the United States.

   A. one million
   B. two millions
   C. 500,000
   D. 150,000

39. Despite public concerns about pornography, _____ think pornography should be protected by the constitutional guarantees of free speech and a free press.

   A. one-third
   B. two-thirds
   C. three-fourths
   D. one half

40. Which country has the highest teen pregnancy among developed nations?

   A. United States
   B. Czechoslovakia
   C. Hungary
   D. New Zealand

41. According to the text, teens account for almost _____ percent of all births in the United States.

   A. 50
   B. 13
   C. 31
   D. 9

42. A recent trend in teen mothers is:

   A. involvement in prostitution
   B. they overcome poverty
   C. they will become single parents, forgoing marriage
   D. effective use of contraceptives

43. Why do so many teens become pregnant?

   A. peer pressure
   B. their parents want to be grandparents
   C. ineffective use of birth control
   D. to collect a welfare check

44. According to the text, a sexually active teenager not using contraceptives has a _____ percent chance of becoming pregnant within the first year of sexual intercourse.

    A. 90
    B. 80
    C. 70
    D. 60

45. According to the text, most researchers interpret incest as resulting from the intersection of _____ and gender within family structures.

    A. sex
    B. age
    C. power
    D. none of the above

46. Teen pregnancy is integrally linked to:

    A. the gender expectations of men and women in society
    B. the lack of government programs
    C. the amount of sexuality revealed on television
    D. the lower socioeconomic strata of society

47. A form of sexual coercion which is common on college campuses is:

    A. fraternity parties
    B. date rape
    C. co-ed swimming classes
    D. taking a Sociology 101 course

48. What race of women are more aware of their vulnerability to rape?

    A. White
    B. Hispanic
    C. Oriental
    D. Black

49. According to the text, _____ percent of the public think that gay marriage should be valid.

    A. 63
    B. 31
    C. 55
    D. 42

50. The widespread changes in men's and women's roles and a greater public acceptance of sexuality as a normal part of social development is:

    A. the sexual revolution
    B. sexology
    C. sexual politics
    D. fiction, since no such phenomenon occurred

Chapter 13

True/False Questions:

T   F   1. Human sexual attitudes and behavior are constant from one culture to another culture.

T   F   2. Sexual attitudes and behavior change over time.

T   F   3. Sexual socialization begins at puberty.

T   F   4. The economic institutions of society are not influence by sex.

T   F   5. Sexology refers to the widespread changes in men's and women's roles in society.

T   F   6. Sigmund Freud presented a developmental model of sexuality originating in childhood and developing over the life cycle.

T   F   7. Havelock Ellis was an early sexologist (1859-1939) who saw sexual dysfunction as the result of psychological problems, rather than organic problems.

T   F   8. Masters and Johnson began their studies using virgins as their sex research subjects.

T   F   9. Young people are becoming sexually active much later in their adolescence.

T   F   10. Having only one sex partner in one's lifetime is the societal standard today.

T   F   11. Functionalism is a microsociological theory and therefore focuses on sexual identities.

T   F   12. Sexual politics refers to the link feminists argue exists between sexuality and power.

T   F   13. Sexual orientation is how individuals experience sexual arousal and pleasure.

T   F   14. In addition to gender, sexual politics are integrally tied to race and class relations in society.

T   F   15. Gender expectations emphasize passivity for men and assertiveness for women in sexual encounters.

T   F   16. Sociology reveals there is no connection between poor women and women of color and sexual exploitation.

T   F   17. Abortion is one of the most seriously contended political issues in recent years.

T   F   18. Japan has the highest rate of teen pregnancy in the world even though levels of teen sexual activity around the world are roughly comparable.

T   F   19. Today, teen mothers feel less pressure to marry than they did in the past.

T   F   20. *Cybersex* is known as sex via the Internet and refers to mutual online sex.

Short Answer Questions:

1. Discuss how conflict theorists see sexuality.
2. Define sexual politics. Provide examples.
3. Compare and contrast between homophobia and heterosexism.
4. Discuss the sexual revolution.

Chapter 13

ANSWERS FOR CHAPTER THIRTEEN

Multiple Choice Questions:

1. D (p.333)
2. B (p.334)
3. C (p.335)
4. D (p.335)
5. D (pp.335-336)
6. C (p.338)
7. B (p.338)
8. A (p.338)
9. B (p.338)
10. D (p.339)
11. C (p.339)
12. B (p.341)
13. C (p.341)
14. D (p.342, Table 13.1)
15. A (p.342, Table 13.1)
16. B (p.342, Table 13.1)
17. A (p.342, Table 13.1)
18. C (p.343)
19. D (p.343)
20. B (p.344)
21. A (p.344)
22. A (p.345)
23. D (p.346)
24. B (p.346)
25. C (p.347)
26. C (p.347)
27. D (p.348)
28. B (p.349)
29. D (p.349)
30. A (p.350)
31. B (p.350)
32. A (p.351)
33. C (p.351)
34. C (p.351)
35. B (p.351)
36. C (p.351)
37. A (p.352)
38. A (p.352)
39. B (p.352)
40. A (p.352)
41. B (p.352)
42. C (pp.352-353)
43. C (p.353)
44. A (p.353)
45. C (p.354)
46. A (p.354)
47. B (p.354)
48. D (p.355)
49. D (p.355)
50. A (p.356)

True/False Questions:

1. F (p.335)
2. T (p.335)
3. F (p.336)
4. F (p.337)
5. F (p.338)
6. T (p.338)
7. T (p.338)
8. F (p.339)
9. F (p.339)
10. F (p.340)
11. F (p.342)
12. T (p.343)
13. T (p.344)
14. T (p.344)
15. F (p.345)
16. F (p.346)
17. T (p.351)
18. F (p.352)
19. T (p.353)
20. T (p.357)

Short Answer Questions:

1. See p.342
2. See pp.344-345
3. See pp.346-347
4. See pp.356-357

CHAPTER FOURTEEN

AGE AND AGING

Multiple Choice:

1. According to the text, since 1950, the number of those over age fifty in the United States has:

    A. remained the same
    B. doubled
    C. tripled
    D. none of the above

2. Life expectancy is shaped by:

    A. gender
    B. race
    C. social class
    D. all of the above

3. _____ refers to how old one things of oneself as being.

    A. "Ideal age"
    B. "Cognitive age"
    C. "Chronological age"
    D. none of the above

4. In many African nations, social status _____ with age.

    A. is unchanged
    B. decreases
    C. increases
    D. none of the above

5. According to the text, those who are most physically active perceive themselves to be _____ than they actually are.

    A. older
    B. younger
    C. smarter
    D. happier

Chapter 14

6. In Asian societies, the Confucian principle of _____ emphasizes respect for the aged.

    A. *familial obligation*
    B. *filial affirmation*
    C. *familial job*
    D. *filial duty*

7. _____ is the biological cessation of ovulation and the menstrual cycle.

    A. Senescence
    B. Senility
    C. Menopause
    D. none of the above

8. Research shows that _____ influence people's perception of others.

    A. age
    B. gender
    C. both A and B
    D. none of the above

9. According to the text, _____ older people are satisfied with their bodies.

    A. most
    B. all
    C. few
    D. none of the above

10. Studies of menopausal women have shown that:

    A. the majority feel happy about the loss of ovulation ability
    B. the symptoms of menopause depicted in social stereotypes are less widespread than generally believed
    C. menopause is related to serious depression
    D. how a woman experiences menopause depends, in part, on how she adopts to the social attitudes that culturally define menopause

11. About _____ percent of the elderly suffer from Alzheimer's disease.

    A. 5
    B. 10
    C. 25
    D. 30

12. According to the text, a basic sociological idea is that perceptions of aging are:

    A. socially constructed
    B. biological determined
    C. psychologically determined
    D. genetically constructed

13. About _____ percent of all elderly are in a nursing home or other institution at any particular time.

    A. 25
    B. 5
    C. 10
    D. 15

14. Until recently, people were required to retire at age:

    A. sixty-five
    B. sixty
    C. fifty-five
    D. seventy

15. According to the text, you cannot take a seat in the House of Representatives until you are:

    A. thirty
    B. thirty-five
    C. eighteen
    D. twenty-five

16. Among the _____ Indians, mourning was reserved only for those who died in their prime because they were seen as a greater loss to the well-being of the community.

    A. Seminole
    B. Comanche
    C. Sioux
    D. Navajo

17. An aggregate group of people born during the same period is called a(n):

    A. age-based subculture
    B. age strata
    C. age cohort
    D. generational frame

18. America's older population will double by 2030, reaching some:

    A. 12 million
    B. 50 million
    C. 25 million
    D. 70 million

19. In 1900, only _____ percent of the population was over age sixty-five.

    A. 15
    B. 4
    C. 9
    D. 22

Chapter 14

20. The most rapid growth among the older population is occurring among those:

    A. under 64 years of age
    B. 65-74 years of age
    C. 75-84 years of age
    D. 85 years and over

21. Social Security was first established in:

    A. 1925
    B. 1945
    C. 1935
    D. 1965

22. Social Security was first established as:

    A. Old Age, Survivors', and Disability Insurance
    B. Old Age and Survivors' Insurance
    C. Old Age and Disability Insurance
    D. none of the above

23. Which of the following is an example of a *rite of passage*?

    A. a christening
    B. a bar mitzvah
    C. a quinceañera
    D. all of the above

24. Thinking of childhood as a separate stage of life dates back to:

    A. the sixteenth and seventeenth centuries in Europe
    B. the Middle Ages (1200-1400 AD) in Europe
    C. the discovery of the Americas
    D. the U.S. at the end of World War I (1918)

25. Americans seeing children as emotionally priceless, but not necessarily defined as economically useful is referred to as the _____ of children.

    A. sanctification
    B. economic transformation
    C. triviatization
    D. sentimentalization

26. In _____, more than one-fourth of the homeless are children.

    A. Canada
    B. Australia
    C. the United States
    D. Denmark

Age and Aging

27. According to the text, _____ women are the group most likely to have to return to paid employment following retirement.

    A. Hispanic
    B. African American
    C. White, non-Hispanic
    D. Asian American

28. In the 1930s, Social Security defined men as _____ and women as _____.

    A. dependents; wage earners
    B. wage earners; dependents
    C. full-time workers; part-time workers
    D. part-time workers; full-time workers

29. In the calculation of Social Security benefits, women who have left the labor force to care of children are:

    A. disadvantaged
    B. advantaged
    C. treated the same as women who did not leave the labor force
    D. none of the above

30. Studies have found that feelings of optimism actually _____ with age.

    A. are unchanged
    B. decrease
    C. increase
    D. none of the above

31. Studies of those who are widowed have revealed that:

    A. five times more men are widowed than women
    B. women tend to suffer more from loneliness after being widowed than men
    C. as widows, women participate in more organizations than do widowed men
    D. all of the above are true

32. Family members provide _____ to _____ percent of long-term care for the elderly.

    A. 20; 30
    B. 60; 70
    C. 40; 50
    D. 80; 90

33. According to the text, the average annual cost of a nursing home is now about _____ per person per year.

    A. $31,000
    B. $23,000
    C. $57,000
    D. $45,000

Chapter 14

34. Medicare and Medicaid are the two _____ programs that assist the elderly with health care and the cost of living.

   A. state
   B. federal
   C. local
   D. none of the above

35. Which of the following is most likely to be engaged in direct physical elder abuse?

   A. daughters-in-law
   B. daughters
   C. sons
   D. sons-in-law

36. The average cost of a basic funeral is:

   A. $5,000
   B. $1,000
   C. $15,000
   D. $7,000

37. _____ is the act of killing a severely ill person or allowing the person to die as an act of mercy.

   A. Requested-death
   B. Euthanasia
   C. Secular death
   D. none of the above

38. The physician-assisted suicide movement is also called:

   A. "shrouding movement"
   B. "cooling board movement"
   C. "requested death movement"
   D. "managed-death movement"

39. _____ percent of the U.S. public say that physician-assisted suicide is morally wrong.

   A. Fourteen
   B. Fifty-five
   C. Thirty
   D. Forty-nine

40. While age prejudice is a(n) _____, age discrimination involves _____.

   A. attitude; behavior
   B. behavior; attitude
   C. assessment; principle
   D. principle; assessment

Age and Aging

41. According to the text, which of the following groups is most likely to vote?

    A. college students
    B. the elderly
    C. young parents
    D. racial/ethnic minorities

42. The Age Discrimination Employment Act was first passed in:

    A. 1967
    B. 1972
    C. 1954
    D. 1978

43. Poverty among the aged is more pronounced for _____ than _____.

    A. women; men
    B. Latinos; Whites
    C. Native Americans; Whites
    D. all of the above

44. Elderly _____ have less access to doctors, hospitals, and other health care facilities than any other social or ethnic group.

    A. Whites
    B. Latinos
    C. African Americans
    D. Native Americans

45. According to the text, even travel time to health care providers is substantially longer for _____ than for any other group.

    A. Whites
    B. Native Americans
    C. African Americans
    D. Latinos

46. According to _____, the elderly voluntarily withdraw from society by retiring and lessening their participation in social activities.

    A. symbolic interaction theory
    B. conflict theory
    C. feminist theory
    D. functionalism

Chapter 14

47. _____ argues that barring youth and the elderly from the labor market is a way of eliminating these groups from competition.

   A. Social construction theory
   B. Symbolic interactionism
   C. Functionalism
   D. Conflict theory

48. _____ considers the role of social meanings in understanding the sociology of age.

   A. Social exchange theory
   B. Functionalism
   C. Conflict theory
   D. Symbolic interactionism

49. _____ analyzes how people explain and manage death.

   A. Conflict theory
   B. Functionalism
   C. Social exchange theory
   D. Symbolic interactionism

50. According to _____, the diminished usefulness of the elderly justifies their depressed earning power.

   A. labeling theory
   B. symbolic interactionism
   C. conflict theory
   D. functionalism

True/False Questions:

T   F   1.   The process of aging involves numerous factors - physical, psychological, and social.

T   F   2.   Most old people have no interest or capacity for sex.

T   F   3.   The percentage of characters on prime-time TV who are presumably over age sixty-five increased between 1975 and 1990.

T   F   4.   Other people's definitions of aging do not affect the way we think about aging.

T   F   5.   Age differentiation focuses on the division of labor or roles in a society on the basis of age.

T   F   6.   Age stratification is the hierarchical ranking of age groups.

T   F   7.   By the year 2025, over 52 percent of the population will be age 65 or over.

Age and Aging

T   F   8.   By 2050, 84 percent of the elderly population will be African American or Hispanic.

T   F   9.   Changes in family structure further alter the traditional patterns of intergenerational care.

T   F   10.   Elderly widowed men are four times more likely to commit suicide than married men in the same age bracket.

T   F   11.   Medicaid and Medicare pay for all expenses in nursing home care.

T   F   12.   Physical and mental abuse of the elderly has been well documented since 1776.

T   F   13.   The funeral home industry is $16 billion business.

T   F   14.   The hospice movement developed as an alternative to hospital-based, technologically controlled death.

T   F   15.   Unfortunately, there are no existing civil right groups that are advocates for the elderly and their issues.

T   F   16.   Quadruple jeopardy is a phrase referring to the simultaneous effects of being old, minority, female, and poor.

T   F   17.   In 2002, more than half of American Indian elderly were below the poverty level.

T   F   18.   In industrialized societies, advances in medicine and technology tend to result in longer life expectancy.

T   F   19.   Disengagement theory predicts that as people age, they gradually withdraw from participation in society.

T   F   20.   Conflict theory assesses the different meaning attributed to social entities, like different age groups.

Short Answer Questions:

1. Compare and contrast between cognitive and chronological age.
2. What are age stereotypes? Provide examples.
3. Define what is meant by generational equity.
4. Compare and contrast between Medicare and Medicaid.

Chapter 14

ANSWER FOR CHAPTER FOURTEEN

Multiple Choice Questions:

1. B (p.363)
2. D (p.365)
3. B (p.365)
4. C (p.365)
5. B (p.365
6. D (p.366)
7. C (p.366)
8. C (p.366)
9. A (p.366)
10. A (p.366)
11. A (p.367, Table 14.1)
12. A (p.367)
13. B (p.367, Table 14.1)
14. A (p.368)
15. D (p.369)
16. B (p.369)
17. C (p.369)
18. D (p.370)
19. B (p.370)
20. D (p.370)
21. C (p.371)
22. A (p.371)
23. D (p.373)
24. A (p.374)
25. D (p.374)
26. C (p.374)
27. B (p.377)
28. B (p.377)
29. A (p.377)
30. C (p.378)
31. B (p.378)
32. D (pp.378-379)
33. C (p.380)
34. B (p.380)
35. C (p.381)
36. A (p.381)
37. B (p.383)
38. C (p.383)
39. D (p.383)
40. A (p.384)
41. B (p.384)
42. A (p.384)
43. D (p.385)
44. D (p.385)
45. B (p.385)
46. D (p.387)
47. D (p.387)
48. D (p.387)
49. D (p.387)
50. D (p.387)

True/False Questions:

1. T (p.364)
2. F (p.367, Table 14.1)
3. F (p.367)
4. F (p.367)
5. T (pp.368-369)
6. T (p.369)
7. F (p.370)
8. F (p.370)
9. T (p.371)
10. T (p.378)
11. F (pp.380-381)
12. F (p.381)
13. T (p.381)
14. T (p.383)
15. F (p.384)
16. T (p.385)
17. T (p.385)
18. T (p.386)
19. T (p.387)
20. F (p.387)

Short Answer Questions:

1. See p.365
2. See pp.366-367
3. See pp.371-373
4. See pp.380-381

CHAPTER FIFTEEN

FAMILIES

Multiple Choice:

1. Sociologists address the family as:

    A. psychological structures
    B. a biological foundation
    C. genetic markers
    D. a social institution

2. Our social institutions presume that families will be:

    A. homosexual
    B. heterosexual
    C. polygamous
    D. none of the above

3. A primary group of people, usually related by ancestry, marriage, or adoption, who form a cooperative economic unit to care for offspring (and each other) and who are committed to maintaining the group over time is a(n):

    A. kinship group
    B. household
    C. extended family
    D. family

4. Traditionally, the family has been defined as a social unit of people related through:

    A. marriage
    B. birth
    C. adoption
    D. all of the above

5. _____ is most commonly associated with Mormons.

    A. Polyandry
    B. Serial monogamy
    C. Exogamy
    D. Polygyny

151

Chapter 15

6. According to the U.S. Census Bureau, all persons occupying a housing unit who may or may not be related comprise a(n):

    A. extended family
    B. household
    C. family
    D. kinship group

7. Many cultures have a tradition of _____ marriages, in which parents make rationally calculated choices about the appropriate marriage partner for their children.

    A. arranged
    B. neolocal
    C. bilateral
    D. selective

8. Today, _____ percent of White Americans disapprove of interracial marriage.

    A. 73
    B. 17
    C. 33
    D. 24

9. Interracial couples are now about _____ percent of married couples in the United States.

    A. 2
    B. 17
    C. 24
    D. 33

10. According to the text, California enacted a law in _____ prohibiting marriage between a White person and any "negro, mulatto, or Mongolian."

    A. 1892
    B. 1967
    C. 1934
    D. 1880

11. Societies in which men and women share power equally are termed:

    A. exogamy
    B. endogamy
    C. egalitarian
    D. no term is available, since such societies do not exist

12. Studies have found that traditionally in _____ societies, the activity of both men and women was culturally defined as critical to economic survival.

    A. most Muslim
    B. Eskimo
    C. newly industrialized
    D. all of the above

13. Families in which there are a large group of related kin in addition to parents and children living together within the same household are called _____ families.

    A. nuclear
    B. conjugal
    C. extended
    D. stem

14. The Chinese Exclusion Act of _____ prohibited the wives and children of resident Chinese laborers from entering the country.

    A. 1851
    B. 1924
    C. 1890
    D. 1882

15. As a result of the Chinese Exclusion Act, there were _____ Chinese males for every female in 1890.

    A. 10.6
    B. 26.8
    C. 13.7
    D. 35.8

16. According to _____, marriage is a mutually beneficial exchange wherein women receive protection, economic support, and status.

    A. symbolic interactionism
    B. conflict theory
    C. feminist theory
    D. functionalism

17. According to _____, in traditional marriages, men get the services that women provide, thus meeting societal needs.

    A. conflict theory
    B. symbolic interactionism
    C. functionalism
    D. feminist theory

Chapter 15

18. According to _____, over time, other institutions have begun to take on some functions originally performed solely by the family.

   A. functionalists
   B. conflict theorists
   C. symbolic interactionists
   D. social constructionists

19. According to _____, the rising numbers of female-headed and single-parent households are the result of social disorganization.

   A. social constructionists
   B. conflict theorists
   C. symbolic interactionists
   D. functionalists

20. Which of the following interprets the family as a system of power relations?

   A. conflict theory
   B. functionalism
   C. symbolic interactionism
   D. social construction theory

21. According to _____, families serve capitalism.

   A. functionalists
   B. symbolic interactionists
   C. conflict theorists
   D. social constructionists

22. _____ conceptualizes the family as an integrative institution.

   A. Conflict theory
   B. Functionalism
   C. Symbolic interactionism
   D. Social construction theory

23. Feminists have been critical of _____ theory for assuming an inevitable gender division of labor within the family.

   A. functionalist
   B. conflict
   C. symbolic interactionist
   D. social construction

Families

24. According to _____, when two people get married, they form a new relationship that has a specific meaning within society.

   A. conflict theorists
   B. functionalists
   C. symbolic interactionists
   D. social exchange theorists

25. _____ emphasizes the construction of meaning within families.

   A. Symbolic interaction theory
   B. Conflict theory
   C. Functionalism
   D. Social exchange theory

26. _____ theorists study how different people define and understand their family experience and how people negotiate family relationships.

   A. Symbolic interaction
   B. Functionalist
   C. Conflict
   D. Feminist

27. _____, once the major cause of early family disruption, has been replaced by _____.

   A. Paid work; household chores
   B. Household chores; paid work
   C. Divorce; death
   D. Death; divorce

28. _____ is the second reason for the increase in the number of female-headed households.

   A. Decreasing number of marriages
   B. Widowhood
   C. Divorce
   D. none of the above

29. Following divorce, of those supposed to receive child support payments, only _____ actually receive any money.

   A. one-quarter
   B. two-thirds
   C. half
   D. none of the above

155

Chapter 15

30. The average amount of child support paid in a year is:

   A. $7377
   B. $5388
   C. $3787
   D. $1377

31. Researchers have found that gay and lesbian couples tend to be less _____ in their household roles than heterosexual couples.

   A. flexible
   B. adaptable
   C. variable
   D. gender-stereotyped

32. Some researchers have concluded that, regardless of their sexual orientation, men have been socialized to believe that:

   A. money equals power
   B. men should serve women
   C. children will make them complete
   D. none of the above

33. Children raised in gay or lesbian families are _____ likely than children raised in heterosexual families to become gay themselves.

   A. less
   B. no more
   C. more
   D. none of the above

34. Never married people currently comprise _____ percent of the U.S. population.

   A. 12
   B. 28
   C. 36
   D. 41

35. According to the text, "hooking up" commonly occurs when both participants are:

   A. drinking
   B. drunk
   C. both A and B
   D. sober

Families

36. _____, meaning sexual interaction without commitment, is widespread on college campuses.

   A. "Going out"
   B. "Clinching"
   C. "Being bad"
   D. "Hooking up"

37. According to the text, _____ percent of people say it does not matter which parent remains the full-time worker.

   A. 45
   B. 69
   C. 38
   D. 23

38. Among couples where both partners are employed, only _____ percent share the housework equally.

   A. 9
   B. 35
   C. 14
   D. 28

39. According to the text, with regard to race, _____ husbands provide a greater share of housework than do White husbands.

   A. African American
   B. Latino
   C. Native American
   D. Asian American

40. According to the text, divorce is _____ likely for couples who marry young.

   A. least
   B. not
   C. less
   D. more

41. The _____ approach emphasizes that violence occurs in families because the society condones violence.

   A. *family cycle*
   B. *family violence*
   C. *family pathology*
   D. none of the above

157

Chapter 15

42. The National Violence Against Women Office estimates that _____ percent of women will be raped, physically assaulted, or stalked by an intimate partner in their lifetime.

    A. 45
    B. 35
    C. 25
    D. 15

43. Which is not true for the feminist approach to family violence?

    A. it argues that since most of the violence in the family is directed against women, the imbalance of power between men and women in the family is the source of most domestic violence
    B. it emphasizes the degree to which many women are trapped in violent relationships.
    C. it emphasizes that violence occurs in families because society condones it
    D. all the above

44. A history of incest has been related to a variety of other problems including:

    A. drug abuse
    B. delinquency
    C. runaways
    D. all of the above

45. According to the text, one in _____ Filipinos directly depends on migrant workers' earnings.

    A. three
    B. ten
    C. two
    D. five

46. The concept of "care work" includes such things as:

    A. child care
    B. elder care
    C. self care
    D. all of the above

47. According to the text, under the Family and Medical Leave Act, employees must have been employed by the granting employer at least _____ to be eligible.

    A. six months
    B. one year
    C. two years
    D. five years

48. Currently, only _____ percent of U.S. workers have child-care benefits available to them from employers.

    A. 45
    B. 6
    C. 13
    D. 28

49. Single-parent families need _____ percent of their income to pay for child care.

    A. 6
    B. 36
    C. 16
    D. 26

50. _____ of children age three in the United States now spend much of their time in child-care centers.

    A. Half
    B. A quarter
    C. Seventy-five percent
    D. Over ninety percent

True/False Questions:

T   F   1.   Sociologists address the family as a social problem.

T   F   2.   The U.S. Census Bureau defines a household as all persons occupying a housing unit who may or may not be related.

T   F   3.   Polygamy is the practice of a woman having more than one husband.

T   F   4.   Exogamy is the practice of selecting mates from within one's group.

T   F   5.   Miscegenation refers to laws that prohibit interracial marriages.

T   F   6.   Matrilineal kinship systems are those in which ancestry is traced through the mother.

T   F   7.   Egalitarian societies are those where men and women share power equally.

T   F   8.   According to the functionalist theory, families are gendered institutions that reflect the gender hierarchies in society.

T   F   9.   Symbolic interactionists see the married relationship as socially constructed.

T   F   10.   Families are systems of social relationships that emerge in response to social conditions and that, in turn, shape the future direction of society.

Chapter 15

T	F	11.	Regardless of race, teen mothers are among the most disadvantaged groups in society.
T	F	12.	The majority of single-parent families are headed by men.
T	F	13.	Commuter marriages are an arrangement when one partner in a dual-earner couple relationship resides in a different city much of the time.
T	F	14.	More than three times as many couples are living together without being married now than was true in the 1970s.
T	F	15.	The rate of divorce continues to increase in the United States.
T	F	16.	Divorce is somewhat higher among African Americans than among Whites.
T	F	17.	Compared to men, women who experience violence are twice as likely to be injured.
T	F	18.	One of the major global changes for the family, especially in the U.S., is geographic mobility.
T	F	19.	The family is the only social institution that typically takes the blame for all of society's problems.
T	F	20.	Child-care workers are some of the highest paid workers in the United States.

Short Answer Questions:

1. Compare and contrast between exogamy and endogamy.
2. List and discuss at least three societal needs filled by the family.
3. What is a "commuter marriage"? Provide examples.
4. Describe the *feminist approach* to partner violence.

ANSWER FOR CHAPTER FIFTEEN

Multiple Choice Questions:

1. D (p.392)
2. B (p.392)
3. D (p.393)
4. D (p.393)
5. D (p.393)
6. B (p.393)
7. A (p.394)
8. C (p.394)
9. A (p.394)
10. D (p.395)
11. C (p.396)
12. B (p.396)
13. C (p.396)
14. D (p.398)
15. B (p.398)
16. D (p.398)
17. C (p.398)
18. A (p.399)
19. D (p.399)
20. A (p.399)
21. C (p.399)
22. B (p.399)
23. A (pp.399-400)
24. C (p.400)
25. A (p.400)
26. A (p.400)
27. D (p.401)
28. C (p.402)
29. B (p.403)
30. C (p.403)
31. D (p.405)
32. A (p.405)
33. B (p.406)
34. B (p.406)
35. C (p.407)
36. D (p.407)
37. B (p.408)
38. D (p.409)
39. A (p.409)
40. D (p.411)
41. B (p.412)
42. C (p.412)
43. C (p.412)
44. D (p.413)
45. D (p.413)
46. D (p.415)
47. B (p.416)
48. C (p.417)
49. C (p.417)
50. A (p.417)

True/False Questions:

1. F (pp.392-393)
2. T (p.393)
3. F (p.393)
4. F (p.394)
5. F (p.394)
6. T (p.395)
7. T (p.396)
8. F (p.398)
9. T (p.400)
10. T (p.401)
11. T (p.402)
12. F (p.403)
13. T (p.404)
14. T (p.408)
15. F (p.410)
16. T (p.411)
17. T (p.412)
18. T (p.414)
19. T (p.415)
20. F (p.417)

Short Answer Questions:

1. See p.394
2. See pp.398-399
3. See p.404
4. See p.412

Chapter 16

CHAPTER SIXTEEN

EDUCATION

Multiple Choice:

1. By _____, compulsory education was established by law in all states, excluding a few Southern states.

    A. 1930
    B. 1960
    C. 1910
    D. 1900

2. In 1910, less than _____ percent of White eighteen-year-olds in the U.S. graduated from high school.

    A. 40
    B. 30
    C. 20
    D. 10

3. Since 1960, attendance in both high school and college has:

    A. decreased
    B. increased
    C. remained the same
    D. none of the above

4. The high school graduation rate for Whites has increased steadily to almost _____ percent in 2003.

    A. 88
    B. 90
    C. 68
    D. 58

5. According to the text, modern industrialized societies need a system that trains people for:

    A. jobs
    B. graduate school
    C. higher learning
    D. none of the above

162

Education

6. According to the text, the higher the educational attainment of a person, the more likely that person will be:

   A. upper class
   B. White
   C. male
   D. all of the above

7. _____ argue that educational level is a mechanism for producing and reproducing inequality in our society.

   A. Symbolic interactionists
   B. Functionalists
   C. Conflict theorists
   D. none of the above

8. _____ focuses on what arises from the operation of the interaction process during the schooling experience.

   A. Functionalism
   B. Symbolic interaction
   C. Conflict theory
   D. Feminist theory

9. According to sociologists, the indicators of socioeconomic status include:

   A. amount of schooling
   B. income
   C. occupation
   D. all of the above

10. A man with no graduate education but a college-only education earns _____ than a woman with a master's degree.

    A. the same
    B. less
    C. more
    D. none of the above

11. According to the text, men with professional degrees (law, medicine, and so forth) earn a median annual income of:

    A. $45,999
    B. $49,180
    C. $81,606
    D. $33,035

Chapter 16

12. According to the text, there is _____ relationship between education, occupation, and income.

   A. no straightforward
   B. direct
   C. absolutely no
   D. none of the above

13. The SAT has an effect in the United States similar to that of the _____ in England, directing the futures of the young according to the results of widely administered exams.

   A. Eleven Plus
   B. Oxford standardized exam
   C. Cambridge standardized exam
   D. *Abitur*

14. According to the text, each _____ increase in family income is worth about 10 to 15 more points on either the SAT verbal of the SAT math tests.

   A. $20,000
   B. $5,000
   C. $1,000
   D. $10,000

15. _____, as a group, have scored higher than Whites in recent years on the quantitative sections of the SAT but somewhat lower on the verbal sections.

   A. Asian Americans
   B. Hispanics
   C. African Americans
   D. Native Americans

16. The *Abitur* is administered in:

   A. England
   B. Mexico
   C. Germany
   D. France

17. Which country's educational system offers more social mobility?

   A. Japan
   B. Germany
   C. United States
   D. Afghanistan

18. The SAT is an example of a(n):

    A. achievement test
    B. standardized ability test
    C. crystallized memory test
    D. none of the above

19. Since early in the twentieth century, educators in our society from preschools to universities have attempted to measure intelligence by means of:

    A. multidimensional tests
    B. unidimensional tests
    C. standardized tests
    D. credentialized tests

20. Which of the following is an example of an achievement test?

    A. Advanced placement (AP) exams
    B. SAT
    C. ACT
    D. PSAT

21. High school grades are _____ the SATs in predicting college grades.

    A. about as accurate as
    B. more accurate than
    C. less accurate than
    D. none of the above

22. Herrnstein and Murray estimate that intelligence is about _____ percent genetically heritable.

    A. 90
    B. 70
    C. 30
    D. 50

23. Herrnstein and Murray estimate that intelligence is about _____ percent the result of social environment.

    A. 90
    B. 70
    C. 50
    D. 30

Chapter 16

24. Some studies show that the similarity in intelligence among truly separated identical twins is only about _____ percent.

    A. 20
    B. 50
    C. 30
    D. 40

25. According to the authors of *The Bell Curve*, the upper and upper-middle classes constitute a genetically-based _____ in America, consisting of those with high IQs, high incomes and prestigious jobs.

    A. credentialed elite
    B. cognitive elite
    C. multidimensional intelligentsia
    D. elite baby-boomer generation

26. According to the text, men have tended to score higher than women in:

    A. numerical reasoning
    B. spatial perception
    C. mechanical aptitude
    D. all of the above

27. According to the text, women have tended to score higher than men in:

    A. perception of detail
    B. memory
    C. certain verbal skills
    D. all of the above

28. Advocates of _____ point out that students in the lower tracks get less teacher attention and simply learn less.

    A. tracking
    B. ability-grouping
    C. detracking
    D. cognitive segregation

29. Among the criticisms of the conclusions of the authors of *The Bell Curve* are which of the following?

    A. the authors place too much emphasis on environmental influences on intelligence
    B. the authors drew too many conclusions about between-group differences from within-group results
    C. they place too much emphasis on the argument that intelligence tests and standardized ability tests are not as accurate for some groups as for others
    D. all of the above

Education

30. The basic idea behind tracking is that students will get a better education and be better prepared for life after high school if they are grouped early according to:

    A. mechanical aptitude
    B. spatial perception
    C. numerical reasoning
    D. cognitive ability

31. The effect of the teacher's expectations on student's actual performance, independent of the student's ability, is the:

    A. teacher stereotyping effect
    B. self-fulfilling prophecy
    C. teacher interaction effect
    D. teacher expectancy effect

32. According to the text, students with the same test scores often get assigned to different tracks because of differences in their:

    A. eye color
    B. social class
    C. race
    D. both B and C

33. Insights into the teacher expectancy effect come from _____ theory.

    A. functionalist
    B. conflict
    C. feminist
    D. symbolic interactionist

34. The *stereotype threat effect* is associated with the work of _____ and associates.

    A. Imani Nikitah
    B. Joe Feagin
    C. W. I. Thomas
    D. Claude M. Steele

35. *The Agony of Education* (1996) by Feagin, Hernan, and Nikitah provides useful insights into:

    A. women attending what have traditionally been all-male colleges and universities (e.g., Virginia Military institute, the U.S. Naval Academy, and West Point)
    B. the experience of Black students on White campuses
    C. the tremendous adjustments that foreign students need to make when attending American higher education institutions
    D. the pressures faced by students attending law school

Chapter 16

36. Recent evidence shows that an internalized _____ stereotype can increase one's SAT test performance.

   A. superior
   B. negative
   C. positive
   D. inferior

37. Some states such as _____ and Minnesota spend considerably more per pupil on education that states such as Tennessee.

   A. Mississippi
   B. Louisiana
   C. Connecticut
   D. none of the above

38. _____ are essentially individual scholarships given to parents that can be used to defray to cost of a child's tuition at any school.

   A. Vouchers
   B. Tickets
   C. Coupons
   D. Tokens

39. According to the text, vouchers may be utilized at _____ schools.

   A. private
   B. public
   C. religious
   D. all of the above

40. One persistent issue for reform in education is the problem of _____ for different school districts within the same city.

   A. unequal funding
   B. unequal curricular requirements
   C. standard curricular requirements
   D. none of the above

41. The educational movement which stresses a return to a traditional curriculum delivered with traditional methods is the:

   A. multiculturalism movement
   B. educational reform movement
   C. back-to-basics movement
   D. credentialism movement

Education

42. In _____ schools, public taxes remain the financier of the school but the responsibility for the running of the school and making policy decisions is delegated to the private sector.

   A. private
   B. charter
   C. public
   D. none of the above

43. Some estimate that over _____ students are home-schooled each year.

   A. ten million
   B. two million
   C. five million
   D. a million

44. According to the text, parental dissatisfaction with public schooling has within the last ten to fifteen years led increasing interest in:

   A. parochial schooling
   B. private schooling
   C. home schooling
   D. none of the above

45. The movement toward _____ schools is a form of *privatization*.

   A. voucher
   B. charter
   C. deed
   D. none of the above

46. The push to introduce into elementary, high school, and college curricula more courses on different and diverse subcultures and groups and gender studies is the:

   A. multiculturalism movement
   B. back-to-basics movement
   C. educational reform movement
   D. credentialism movement

47. The Elementary and Secondary Education Act was passed in:

   A. 1994
   B. 1965
   C. 1974
   D. 1988

Chapter 16

48. The *No Child Left Behind Act* was signed into law on January 8:

    A. 2002
    B. 2000
    C. 1998
    D. 2004

49. Which of the following U.S. Presidents proposed a voluntary nationwide program of reading and mathematics testing?

    A. George W. Bush
    B. Jimmy Carter
    C. William Clinton
    D. Ronald Reagan

50. The *No Child Left Behind Act* was put into place as an attempt to close the achievement gap between advantaged and disadvantaged students by means of:

    A. increased funding to schools in poorer areas
    B. improvement in teacher qualifications
    C. improved methods for testing students
    D. all of the above

True/False Questions:

T   F   1. All known societies have some type of education institutions of some sort.

T   F   2. Social control is not a function of education.

T   F   3. Credentialism focuses on the expectations a teacher has for a student's degree program.

T   F   4. The connection between income and education is independent of gender.

T   F   5. In the United States education has never been seen as a means to upward social mobility.

T   F   6. In Germany, the *Abitur* test measures if a student meets the academic standing to enter a university.

T   F   7. Education has reduced many inequalities in society since the turn of the twentieth century.

T   F   8. Cognitive ability has been gauged according to the results of multiple choice tests.

T   F   9. Women are catching up to men on the math part of the SATs.

T   F   10.   The cognitive elite are those people with high IQs, high incomes, and prestigious jobs.

T   F   11.   Advocates of detracking believe that students of varying cognitive abilities should not be in a mixed classroom.

T   F   12.   Self-fulfilling prophecy has no real or valid application to the education environment.

T   F   13.   Gender affects teacher expectations and student performance.

T   F   14.   Standardized tests in math tend to underpredict women's actual grades in mathematics.

T   F   15.   The *stereotype threat effect* is associated with the work of Joe Feagin.

T   F   16.   The back to basics movement in education is a demand to get back to the three R's: reading, writing, and arithmetic.

T   F   17.   Some estimate that over ten million students are home-schooled each year.

T   F   18.   Many states now require teachers to attain a minimum score on a standardized test called the National Teacher Examination (NTE).

T   F   19.   The more than 500 Women's Studies programs across the country are part of the multiculturalism movement.

T   F   20.   The "digital divide" in academic technology has to do with the divide between rural and urban areas.

Short Answer Questions:

1. What are the latent functions of education? Provide examples.
2. Compare and contrast between the symbolic interactionist and the conflict perspectives on education.
3. Who are the *cognitive elite*?
4. Compare and contrast between the back-to-basics movement and the multiculturalism movement.

Chapter 16

ANSWER FOR CHAPTER SIXTEEN

Multiple Choice Questions:

1. D (p.424)
2. D (p.424)
3. B (p.424)
4. B (p.425)
5. A (p.425)
6. D (p.426)
7. C (p.426)
8. B (p.427)
9. D (p.427)
10. C (p.428)
11. C (p.428)
12. A (p.428)
13. A (p.429)
14. D (p.429)
15. A (p.429)
16. C (p.430)
17. C (p.430)
18. B (p.431)
19. C (p.431)
20. A (p.432)
21. A (p.432)
22. B (p.433)
23. D (p.433)
24. B (p.433)
25. B (p.433)
26. D (p.433)
27. D (p.433)
28. C (p.434)
29. D (p.434)
30. D (p.434)
31. D (p.435)
32. D (p.435)
33. D (p.436)
34. D (p.438)
35. B (p.438)
36. C (p.439)
37. C (p.440)
38. A (p.440)
39. D (p.440)
40. A (p.440)
41. C (pp.440-441)
42. B (p.441)
43. D (p.441)
44. C (p.441)
45. B (p.441)
46. A (p.441)
47. B (p.442)
48. A (p.442)
49. C (p.442)
50. D (p.442)

True/False Questions:

1. T (p.425)
2. F (p.425)
3. F (p.426)
4. F (p.428)
5. F (p.428)
6. T (p.430)
7. T (pp.430-431)
8. F (pp.431-432)
9. T (p.433)
10. T (p.433)
11. F (p.434)
12. F (p.436)
13. T (p.436)
14. T (p.437)
15. F (p.438)
16. T (pp.440-441)
17. F (p.441)
18. T (p.441)
19. T (p.441)
20. F (p.443)

Short Answer Questions:

1. See p.425
2. See pp.426-427
3. See pp.433-434
4. See pp.440-441

Religion

CHAPTER SEVENTEEN

RELIGION

Multiple Choice:

1. Which of the following statements about religion is (are) true?

    A. religion is an institutionalized feature of groups
    B. religions are based on beliefs that are considered sacred
    C. religion provides answers to questions of ultimate meaning
    D. all of the above

2. Religion typically establishes _____ for the behavior of believers, some quite strict, such as no sex before marriage.

    A. prescriptions
    B. proscriptions
    C. directives
    D. none of the above

3. According to the text, _____ beliefs often have a supernatural element.

    A. scientific
    B. religious
    C. logical
    D. controlled

4. According to the text, _____ percent of Americans identify themselves as Protestant.

    A. 12
    B. 27
    C. 48
    D. 15

5. According to the text, _____ percent of Americans identify themselves as Catholic.

    A. 48
    B. 37
    C. 18
    D. 25

6. What percentage of Americans identify as Jewish?

    A. 3 percent
    B. 11 percent
    C. 7 percent
    D. 25 percent

173

Chapter 17

7. What percentage of Americans say they attend a church or synagogue weekly?

   A. 14 percent
   B. 54 percent
   C. 34 percent
   D. 64 percent

8. The majority of Americans identify themselves as:

   A. Muslim
   B. Christian
   C. Jewish
   D. no religion

9. The intensity and consistency of a person's or group's faith is:

   A. religiosity
   B. the sacred
   C. religious socialization
   D. polytheism

10. Church membership and attendance are higher among:

    A. women than men
    B. men than women
    C. younger people than older people
    D. Whites than African Americans

11. The number who think that religion can answer all or most of today's problems has been relatively steady at about:

    A. 61 percent
    B. 48 percent
    C. 35 percent
    D. 29 percent

12. Christianity and Judaism are:

    A. polytheistic
    B. monotheistic
    C. secular
    D. matriarchal

13. Which of the following group is most likely to be polytheistic?

    A. White, non-Hispanics
    B. Hispanic Americans
    C. African Americans
    D. Native Americans

14. Religions that are _____, stress interdenominational cooperation and the importance of common religious work

   A. *restricted*
   B. *exclusive*
   C. *ecumenical*
   D. none of the above

15. _____ was the historical source for the ideas relating religion and social conflict.

   A. Karl Marx
   B. Emile Durkheim
   C. Max Weber
   D. Talcott Parsons

16. In _____ view, religious rituals are vehicles for the creation, expression, and reinforcement of social cohesion.

   A. Parsons's
   B. Weber's
   C. Marx's
   D. Durkheim's

17. Which of the following theorists wrote *The Elementary Forms of Religious Life*?

   A. Emile Durkheim
   B. Karl Mark
   C. Max Weber
   D. Talcott Parsons

18. _____ saw religion as a tool for class oppression.

   A. Emile Durkheim
   B. Max Weber
   C. Karl Marx
   D. Talcott Parsons

19. Which theorist stated that religion is an ideology - a belief system that legitimates the social order and supports the ideas of the ruling class?

   A. Max Weber
   B. Emile Durkheim
   C. Karl Marx
   D. George Herbert Mead

Chapter 17

20. According to _____, religious practices and rituals provide definitions of group and individual identity.

   A. functionalism
   B. symbolic interactionism
   C. conflict theory
   D. feminist theory

21. According to _____, religion promotes order by a sense of collective consciousness.

   A. functionalism
   B. symbolic interactionism
   C. conflict theory
   D. feminist theory

22. According to _____, religious practices and rituals define in-groups and out-groups, thus defining group boundaries.

   A. functionalism
   B. symbolic interactionism
   C. conflict theory
   D. feminist theory

23. According to your text, _____ people around the world are Jewish.

   A. 379 million
   B. 14 million
   C. 107 million
   D. 68 million

24. The largest religious group in the United States is:

   A. Jews
   B. Muslims
   C. Protestants
   D. Roman Catholics

25. Christianity developed in the Mediterranean region of:

   A. Europe
   B. Africa
   C. East Asia
   D. South Asia

26. According to the text, there are two main categories of Protestants: *mainline Protestants* and:

   A. *liberals*
   B. *Mormons*
   C. *Catholics*
   D. *conservatives*

27. In the U.S., fundamentalism is the conservative offspring of the _____ movement.

    A. *glossolalia*
    B. faith healing
    C. evangelical
    D. none of the above

28. Roman Catholics identify the _____ as the source of religious authority.

    A. President of Italy
    B. Pope
    C. John Smith
    D. none of the above

29. Which world religion is typically associated with Middle Eastern countries?

    A. Jainism
    B. Buddhism
    C. Islam
    D. Confucianism

30. The teachings of _____ are the source of both Christian beliefs and Islamic beliefs.

    A. Judaism
    B. Buddhism
    C. Hinduism
    D. Confucianism

31. About _____ percent of the world's Jewish population now lives in the United States.

    A. 10
    B. 25
    C. 40
    D. 65

32. The Koran is the holy book of:

    A. Buddhism
    B. Shinto
    C. Judaism
    D. Islam

33. The concepts of *karma* and reincarnation are an important part of _____ religious philosophy.

    A. Islamic
    B. Confucian
    C. Buddhist
    D. Hindu

Chapter 17

34. Which belief system encourages its followers to pursue spiritual transformation with a focus on meditation?

    A. Islam
    B. Buddhism
    C. Confucianism
    D. Judaism

35. Which of the following belief systems is more of a moral code than a sacred religion per se?

    A. Confucianism
    B. Judaism
    C. Islam
    D. Buddhism

36. Hinduism is deeply linked to the social system of:

    A. Iran
    B. India
    C. Sudan
    D. Egypt

37. Confucianism is mostly found in:

    A. Egypt
    B. India
    C. Iran
    D. China

38. About _____ percent of Latinos identify themselves as Catholic.

    A. 90
    B. 50
    C. 60
    D. 70

39. Hasidic Jews and the Amish, who represent examples of extreme forms of exclusive religious groups in the United States, are:

    A. sects
    B. cults
    C. informal religious groups
    D. ethnoreligious groups

40. _____ are religious camps in Pakistan where young boys are taught a strict interpretation of Islam combined with instruction that they should feel threatened by Westerners, especially Americans.

    A. *Karma*
    B. *Berdaches*
    C. *Jihad*
    D. *Madrasas*

41. _____ convey the essential characteristics of some social entity or phenomenon, even though they do not explain every feature of each entity included in the generic category.

    A. Generalizations
    B. Ideal types
    C. Social constructs
    D. Stereotypes

42. _____ are formal organizations that tend to see themselves, and are seen by society, as the primary and legitimate religious institutions.

    A. Cults
    B. Congregations
    C. Sects
    D. Churches

43. _____ socialization occurs when one observes and absorbs the religious perspective of parents and peers.

    A. *Modern religious*
    B. *Traditional religious*
    C. *Formal religious*
    D. *Informal religious*

44. _____ is a transformation of religious identity.

    A. A quest orientation
    B. Conversion
    C. An extreme religious orientation
    D. An intrinsic religious orientation

45. The case of the "American Taliban," John Walker Lindh, is a recent example of:

    A. social programming theory
    B. social drift theory
    C. social disengagement theory
    D. none of the above

Chapter 17

46. Which of the following groups is most likely to identify as "born again"?

   A. men
   B. women
   C. African Americans
   D. both B and C

47. The evangelical movement consists of diverse groups, including:

   A. Faith Assemblies of God
   B. Churches of Christ
   C. Jehovah's Witnesses
   D. all of the above

48. The Christian right sees the changing role of women in society and the influence of the feminist movement as threatening:

   A. national security
   B. the survival of the human race
   C. traditional "family values"
   D. none of the above

49. Two polarized images of women developed within Christianity: woman as temptresses and:

   A. madonnas
   B. priests
   C. leaders
   D. none of the above

50. The process by which society is increasingly organized around rational, empirical, and scientific forms of thought, according to Max Weber, is called:

   A. collective consciousness of society
   B. the McDonaldization of society
   C. the secularization of society
   D. the rationalization of society

True/False Questions:

T   F   1.   Sociologists define religion as a purely private experience.

T   F   2.   The sacred is that which is set apart from ordinary activity and seen as holy, protected by special rites and rituals.

T   F   3.   Religion has nothing to do with establishing norms of behavior for society.

T   F   4.   Over 75 percent of Americans say they belong to a church or synagogue.

T   F   5.   The Untied States is based on the religious tradition of Catholic heritage.

Religion

T   F   6.   Religiosity is the intensity and consistency of practice of a person's/group's faith.

T   F   7.   Polytheism is the worship of more than one deity.

T   F   8.   Exclusive religious groups are those with a moderate and liberal religious orientation.

T   F   9.   Religious rituals are symbolic activities that express a group's spiritual convictions.

T   F   10.   According to Weber, material success along with a strong Protestant work ethic was taken to be a demonstration of favor with God.

T   F   11.   Marx saw religion to be the means to free people from their oppression by the proletariat.

T   F   12.   The largest religion in the world, if measured in terms of numbers of followers, is Islam.

T   F   13.   The United States is one of the most religiously diverse societies in the world.

T   F   14.   Protestants form the largest religious group in the United States.

T   F   15.   One of the amazing thing about Judaism is that compared to other world religions, there are no internal divisions.

T   F   16.   The Buddha in Buddhism is Siddartha Gautama, who as a young man sought the path of enlightenment.

T   F   17.   Most sects tend to hold to very organized and disciplined worship services.

T   F   18.   *Megachurches* have 2000 or more members.

T   F   19.   Religious socialization is the process by which one learns a particular religious faith.

T   F   20.   One central theme of African American religiosity has been liberation from oppression.

Short Answer Questions:

1. Compare and contrast between monotheism and polytheism. Provide examples.
2. What is the Protestant Ethic? Provide examples.
3. Compare and contrast between Hinduism, Buddhism and Confucianism.
4. Compare and contrast between informal religious socialization and formal religious socialization.

Chapter 17

ANSWERS FOR CHAPTER SEVENTEEN

Multiple Choice Questions:

1. D (pp.448-449)
2. B (p.449)
3. B (p.449)
4. C (p.450)
5. D (p.450)
6. A (p.450)
7. C (p.450)
8. B (p.450)
9. A (p.450)
10. A (p.451)
11. A (p.451)
12. B (p.451)
13. D (p.451)
14. C (p.452)
15. A (p.452)
16. D (p.453)
17. A (p.453)
18. C (p.453)
19. C (p.454)
20. B (p.455, Table 17.1)
21. A (p.455, Table 17.1)
22. C (p.455, Table 17.1)
23. B (pp.455-456)
24. C (p.456)
25. A (p.456)
26. D (p.456)
27. C (p.457)
28. B (p.458)
29. C (p.459)
30. A (p.459)
31. C (p.459)
32. D (p.459)
33. D (p.460)
34. B (p.460)
35. A (p.460)
36. B (p.460)
37. D (p.460)
38. D (p.461)
39. D (p.461)
40. D (p.462)
41. B (p.463)
42. D (p.463)
43. D (p.464)
44. B (p.466)
45. B (p.467)
46. D (p.469)
47. D (p.469)
48. C (p.470)
49. A (p.471)
50. D (p.472)

True/False Questions:

1. F (p.448)
2. T (p.448)
3. F (p.449)
4. F (p.450)
5. F (p.450)
6. T (p.450)
7. T (p.451)
8. F (p.452)
9. T (p.453)
10. T (p.453)
11. F (pp.453-454)
12. F (p.455)
13. T (p.456)
14. T (p.456)
15. F (p.459)
16. T (p.460)
17. F (p.463)
18. T (p.463)
19. T (p.464)
20. T (p.470)

Short Answer Questions:

1. See pp.451-452
2. See p.453
3. See p.460
4. See pp.464-465

CHAPTER EIGHTEEN

ECONOMY AND WORK

Multiple Choice:

1. The development of agricultural societies followed the development of technologies that enabled the large-scale production of:

    A. silicon chips
    B. information
    C. garments
    D. food

2. Postindustrial economies are characterized by:

    A. agriculture and factories.
    B. service industries and advanced technology.
    C. a majority of persons employed in manufacturing.
    D. none of the above

3. The _____ Revolution is probably one of the greatest sources of social and economic change in the future.

    A. Agricultural
    B. Steam Revolution
    C. Industrial
    D. Information

4. According to the text, the _____ basis of society in the United States shapes the character of the nation's other institutions.

    A. communist
    B. socialist
    C. capitalist
    D. none of the above

5. _____ philosophy argues that capitalism is fundamentally unjust because powerful owners take more from laborers than they give and use their power to maintain the inequalities between workers and owners.

    A. Industrialist
    B. Socialist
    C. Capitalist
    D. Communist

Chapter 18

6. Which of the following economic systems is characterized by private ownership of the means of production, profit generated by the workers' production of the goods and services, and owners' disproportionate consumption of goods and profits?

   A. capitalism
   B. socialism
   C. communism
   D. none of the above

7. The economic system that is characterized by state ownership and management of the basic industries is:

   A. capitalism
   B. communism
   C. socialism
   D. none of the above

8. Which economic system is described as the state being the sole owner of the systems of production?

   A. capitalism
   B. communism
   C. socialism
   D. none of the above

9. In the Sociology in Practice box "Toys Are NOT U.S.," it is noted that:

   A. the toy industry has created good paying jobs not only in the U.S. but in third world countries as well
   B. the manufacturing of toys is a classic example of the global assembly line
   C. U.S. workers today have an increasing sense of bonding with foreign workers
   D. U.S. foreign investments in other countries have significantly improved working conditions of foreign workers

10. Which of the following statements is true about the global economy?

    A. Research and management are controlled by the most developed countries and assembly-line work is performed in nations with less privileged positions in the global economy
    B. The global economy has had a positive effect on the U.S. economy in particular, creating many new jobs in the manufacturing sector
    C. The real impact of the global economy is on trade, not investments, labor, or technology
    D. The creation of state regulations governing work conditions and terms of employment has made the transfer of jobs overseas less attractive to U.S. manufacturers

11. Recent surveys of the garment industry in Los Angeles found that _____ percent of the garment firms were in violation of health and safety regulations for workers.

    A. 14
    B. 25
    C. 61
    D. 96

12. Which of the following groups is most likely to be unionized?

   A. White men
   B. Black men
   C. Latino men
   D. Asian men

13. The transition from a predominately goods-producing economy to an economy based on the provision of services is:

   A. deindustrialization
   B. job displacement
   C. inflation
   D. economic rebuilding

14. The text points out that in the United States today, at least _____ percent of workers are employed in service industry jobs.

   A. 51
   B. 63
   C. 75
   D. 72

15. According to the text, _____ percent of all workers are union members.

   A. 33
   B. 13
   C. 23
   D. 3

16. Studies of deindustrialization have shown that:

   A. those industries now in decline have been the ones in which the highest number of skilled White workers are employed
   B. White workers are more likely to be fired than Black workers, even when their job seniority, education, experience and performance are the same
   C. among the areas hardest hit are communities that were heavily dependent on a single industry
   D. all of the above are true

17. Electronic sweatshops are:

   A. automated factories where children are chained to the machines
   B. characterize employees whoa re on computers all day
   C. the back offices found in many industries where workers at computer terminals process hundreds or thousands of transactions a day
   D. home computer businesses

Chapter 18

18. As work roles become _____, employees are paid less and have less control over their tasks.

    A. intricate
    B. complex
    C. skilled
    D. deskilled

19. Contingent workers are estimated now to comprise _____ percent of the labor force.

    A. 50
    B. 10
    C. 30
    D. 20

20. According to _____, certain tasks must be done to sustain society, and the organization of work reflects the values and other characteristics of a given social order.

    A. conflict theory
    B. symbolic interactionism
    C. feminist theory
    D. functionalism

21. _____ theorists see class inequality as the source of unequal rewards that workers receive for work.

    A. Conflict
    B. Symbolic interaction
    C. Social exchange
    D. Functionalist

22. _____ theorists are interested in the meaning people give to work.

    A. Symbolic interaction
    B. Conflict
    C. Functionalist
    D. Social exchange

23. *Reproductive labor* is also called:

    A. *labor of love*
    B. *care work*
    C. *filial chores*
    D. none of the above

24. According to the text, in _____, employed women may take eighteen weeks maternity leave, with either parent allowed to take twenty-six extra paid weeks.

  A. France
  B. England
  C. Norway
  D. Italy

25. _____ *labor* involves putting on a false front before clients, and is performed in jobs where inducing or suppressing a feeling in the client is a primary work task.

  A. *Emotional*
  B. *Caring*
  C. *Sympathy*
  D. *Expressive*

26. Which of the following groups is most likely to be employed?

  A. African American men
  B. Native American men
  C. White men
  D. Hispanic men

27. Which of the following groups is least likely to be employed?

  A. African American women
  B. Native American women
  C. Hispanic women
  D. White women

28. In 2003, the labor force in the U.S. included approximately _____ million people.

  A. 214
  B. 146
  C. 100
  D. 178

29. Before the Family and Medical Leave Act of 1993, only _____ percent of medium- to large-sized companies provided maternity and paternity leave.

  A. 12
  B. 27
  C. 31
  D. 44

Chapter 18

30. According to the text, _____ percent of workers work at nonstandard times – weekends, nights, and shifting hours.

   A. 10
   B. 50
   C. 40
   D. 20

31. Studies find large numbers of workers reporting the "_____" of work into their home lives – with work making people too tired to enjoy other parts of life or job worries distracting them when they are at home.

   A. workaholic
   B. mere exposure
   C. take-it-with-you
   D. spillover

32. Which of the following groups earns the most?

   A. White men
   B. White women
   C. Black men
   D. Black women

33. To _____, differential wages are a source of motivation and a means to ensure that the most talented people fill jobs essential to society.

   A. functionalists
   B. symbolic interactionists
   C. conflict theorists
   D. feminists

34. From a _____ perspective, wage inequality maintains race, class, and gender inequality.

   A. functionalist
   B. symbolic interactionist
   C. conflict
   D. social exchange

35. White men and _____ men hold the jobs with the highest occupational prestige.

   A. Asian American
   B. African American
   C. Hispanic American
   D. Native American

Economy and Work

36. The prestige of jobs increases as more _____ enter the field.

   A. women
   B. men
   C. Native Americans
   D. Latinos

37. _____ is defined as the percentage of those not working but looking for work.

   A. The labor non-participant rate
   B. The underemployment rate
   C. The unemployment rate
   D. None of the above

38. The unemployment rate includes:

   A. those who have actively sought to obtain a job during the past four weeks and who is registered with the unemployment office
   B. those who have given up looking for gull-time work
   C. those who earned less than five days pay the week prior to the data being collected
   D. all those who are jobless

39. Manual labor is presumed to be the inverse of mental labor, meaning it is presumed to require:

   A. cognitive ability
   B. high intelligence
   C. superior intellect
   D. no thinking

40. The term _____, originally meaning "bitter strength" and "slave," was used to refer to the Chinese laborers that immigrated to the United States and Hawaii in the mid- and late nineteenth century.

   A. madrasas
   B. *sansei*
   C. *issei*
   D. *coolies*

41. According to the text, almost half of _____ men and women work in the professional and managerial category.

   A. Hispanic American
   B. African American
   C. White American
   D. Asian American

Chapter 18

42. According to the text, _____ are part of the underground economy.

    A. waitresses
    B. machine operators
    C. farm managers
    D. prostitutes

43. According to the text, _____ are part of the secondary labor market.

    A. plumbers
    B. cashiers
    C. janitors
    D. drug dealers

44. According to the text, _____ are part of the primary labor market.

    A. clerical workers
    B. bartenders
    C. cooks
    D. teachers

45. Many service jobs, such as waiting tables, which primarily employ women and minorities, are in:

    A. the underground labor market
    B. the primary labor market
    C. the secondary labor market
    D. the intermediate labor market

46. _____ theorists would argue that the dual labor market favors capitalist owners, who find it advantageous to encourage discord between different groups of workers.

    A. Functionalist
    B. Conflict
    C. Symbolic interaction
    D. Social exchange

47. Sociologist _____ was one of the first to suggest that people with disabilities face issues similar to minority groups.

    A. Talcott Parsons
    B. Irving Zola
    C. Joe Feagin
    D. Lisa Park

48. The Americans with Disabilities Act was adopted by Congress in:

    A. 1972
    B. 1968
    C. 1990
    D. 1989

Economy and Work

49. _____ percent of Americans think gays should be employed in the armed forces and be included in the president's cabinet.

   A. Sixty-three
   B. Seventy-two
   C. Fifty-four
   D. Twenty-four

50. Sexual harassment was first made illegal by _____, which identified sexual harassment as a form of sex discrimination.

   A. Title VII of the Civil Rights Act of 1964
   B. Title IX of the Civil Rights Act of 1954
   C. Title VII of the Civil Rights Act of 1986
   D. Title IX of the Civil Rights Act of 1976

True/False Questions:

T   F   1.   Not all societies are organized around an economic base.

T   F   2.   The economy of a society is the system by which goods and services are produced, distributed, and consumed.

T   F   3.   Socialism is an economic system based on the principles of market competition.

T   F   4.   Capitalism is an economic institution characterized by state ownership and management of the basic industries.

T   F   5.   Communism holds to the pursuit of private property and profit.

T   F   6.   The global economy links the lives of millions of Americans to the experiences of other people throughout the world.

T   F   7.   Economic restructuring refers to the contemporary transformations in the basic structure of work that are permanently altering the workplace.

T   F   8.   Deindustrialization causes job decline in professional and administrative positions with high educational requirements.

T   F   9.   Mismatch theory is an argument that Black's disadvantage in the labor market results from a combination of residential segregation in urban areas and the movement of jobs to suburban areas.

T   F   10. Contingent workers are those who do not hold regular jobs, but whose employment is dependent upon demand.

T   F   11. According to Marx, the work ethic stems from the Protestant belief that hard work is a sign of moral stature, and prosperity is a sign of God's favor.

Chapter 18

T   F   12.   Conflict theory sees work as a necessity which is functionally integrated with other social institutions.

T   F   13.   Conflict theorists see wage inequality as reflecting the differently valued characteristics that workers bring from ensuring that the most talented will fill the most important jobs.

T   F   14.   Occupational prestige has to do with the pattern by which workers are located in the labor force.

T   F   15.   Jobs often lose their prestige as many men a given profession.

T   F   16.   Underemployment is the condition of being employed at a level below what would be expected given a person's training, experience, or education.

T   F   17.   The division of labor is the systematic interrelatedness of different tasks that develops in complex societies.

T   F   18.   The gender division of labor refers to the different work that women and men do in society.

T   F   19.   The racial, class, and gender divisions of labor intersect, creating unique work experiences for different groups.

T   F   20.   Occupations that employ numerous immigrant workers are especially hazardous for employees.

Short Answer Questions:

1. Compare and contrast between socialism and communism.
2. What are *electronic sweatshops*?
3. Compare and contrast between the functionalist and the conflict perspectives on work and the economy.
4. Define sexual harassment. Provide examples.

ANSWERS FOR CHAPTER EIGHTEEN

Multiple Choice Questions:

1. D (p.478)
2. B (p.478)
3. D (pp.478-479)
4. C (p.479)
5. D (p.479)
6. A (p.479)
7. C (p.479)
8. B (p.479)
9. B (p.480)
10. A (p.481)
11. D (p.481)
12. B (p.482)
13. A (p.482)
14. C (p.482)
15. B (p.482)
16. C (p.482)
17. C (p.484)
18. D (p.484)
19. C (p.485)
20. D (p.486)
21. A (p.486)
22. A (p.487)
23. B (p.487)
24. C (p.487)
25. A (p.488)
26. D (p.488)
27. C (p.488)
28. B (p.488)
29. B (p.489)
30. C (p.490)
31. D (p.490)
32. A (p.490)
33. A (p.491)
34. C (p.491)
35. A (p.491)
36. B (p.492)
37. C (p.492)
38. A (p.492)
39. D (p.494)
40. D (p.494)
41. D (p.496)
42. D (p.497, Table 18.4)
43. B (p.497, Table 18.4)
44. D (p.497, Table 18.4)
45. C (p.497)
46. B (p.498)
47. B (pp.498-499)
48. C (p.499)
49. B (p.501)
50. A (p.503)

True/False Questions:

1. F (p.478)
2. T (p.478)
3. F (p.479)
4. F (p.479)
5. F (p.479)
6. T (pp.479-480)
7. T (p.481)
8. F (p.482)
9. T (p.485)
10. T (p.485)
11. F (p.486)
12. F (p.486)
13. F (p.491)
14. F (p.491)
15. F (p.492)
16. T (p.493)
17. T (p.494)
18. T (p.494)
19. T (p.495)
20. T (p.504)

Short Answer Questions:

1. See p.479
2. See p.484
3. See pp.486-487
4. See pp.503-504

Chapter 19

CHAPTER NINETEEN

POLITICS AND GOVERNMENT

Multiple Choice:

1. Which of the following institutions make up the state?

   A. the government
   B. the legal system
   C. the military
   D. all of the above

2. The _____ is the branch of government responsible for defending the nation against domestic and foreign conflicts.

   A. police
   B. prison system
   C. school system
   D. military

3. According to the text, by circulating propaganda and by _____, the state can direct public opinion.

   A. deregulation
   B. censorship
   C. dissemination
   D. none of the above

4. According to the text, the European Community now includes all of the following countries except:

   A. France
   B. Belgium
   C. Turkey
   D. Greece

5. Which of the following countries is(are) part of the European Community?

   A. Luxembourg
   B. Finland
   C. Portugal
   D. all of the above

6. Analysts worry that the increasing similarity produced by global interdependence will produce a _____ where everything will look alike.

   A. *un-culture*
   B. *monoculture*
   C. *non-culture*
   D. none of the above

7. Which of the following countries is not a member of NAFTA?

   A. Costa Rica
   B. Mexico
   C. Canada
   D. the United States

8. _____ is a person's or group's ability to exercise influence and control over others.

   A. Authority
   B. Power
   C. Charisma
   D. none of the above

9. _____ power is achieved through force, often against the will of the people being forced.

   A. Coercive
   B. Legitimate
   C. Renegade
   D. Justifiable

10. _____ authority stems from long-established patterns that give certain people or groups legitimate power in a society.

    A. Charismatic
    B. Traditional
    C. Rational-legal
    D. Expertise-based

11. According to Max Weber, which of the following types of authority leads to the formation of bureaucracies?

    A. traditional authority
    B. charismatic authority
    C. rational-legal authority
    D. all of the above

Chapter 19

12. A German sociologist and an early critic of bureaucracies argued that there is an _____ in bureaucracies.

   A. iron law of coercion
   B. iron law of oligarchy
   C. iron law of rationality
   D. iron law of restraint

13. Robert Michels noted that the formal organization of bureaucracies tends to evolve into a system where _____ become increasingly powerful.

   A. the poor
   B. the people
   C. consumers
   D. a small elite

14. Bureaucracies:

   A. are a type of formal organization characterized by an authority hierarchy, a clear division of labor, explicit rules
   B. only exist in democratic countries
   C. were created and formulated by Italians
   D. All of the above are true of bureaucracies

15. In the _____ model, no particular group is seen as politically dominant.

   A. feminist
   B. autonomous
   C. pluralist
   D. power elite

16. The _____ model of power in society as coming from the representation of diverse interests of different groups in society.

   A. autonomous state model
   B. power-elite
   C. pluralist
   D. feminist

17. Which of the following is not categorized as a theory of power?

   A. oligarchy
   B. power-elite
   C. autonomous state model
   D. feminist

18. According to the text, _____ cannot be given directly to candidates but can be used by political parties for a whole host of party-building expenses.

   A. "soft money"
   B. "money morale"
   C. "elite money"
   D. none of the above

19. The _____ view of the state emphasizes the power of the upper class over the lower classes – the small group of elites over the rest of the population.

   A. structural
   B. exchange
   C. functionalist
   D. Marxist

20. Research on political action committees (PAC's) reveals that:

   A. one reason they are so effective today is that they have existed since the beginning of the twentieth century without any regulation
   B. PAC's serve the legitimate purpose of allowing interest groups to make their opinions felt in a pluralistic system
   C. the influence of PAC's on the political process is greatly exaggerated
   D. in reality, PAC's are fairly equal in resources available to them and the effect that any one of them has on the political system

21. According to the text, the _____ model posits a strong link between government and business.

   A. power elite
   B. pluralist
   C. autonomous
   D. rational-legal

22. Corporations have an intense interest in public policy given that _____ regulates how business is conducted.

   A. the people
   B. market forces
   C. government
   D. none of the above

23. _____ theory sees the state as a network of administrative and policing organization, each with its own interests, such as maintenance of its complex bureaucratic and protection of its special privileges.

   A. Irrational state
   B. Autonomous state
   C. Pluralist-rational
   D. Command state

Chapter 19

24. _____ theory actually builds from the sociological discussion of bureaucracy originally proposed by Max Weber.

   A. Irrational state
   B. Pluralist-rational
   C. Autonomous state
   D. Command state

25. _____ see power as widely dispersed through the class system, while _____ see the state as relatively independent of class interests.

   A. Feminist theorists; power elite theorists
   B. Autonomous state theorists; pluralist theorists
   C. Pluralist theorists; autonomous state theorists
   D. Power elite theorists; feminist theorists

26. _____ see political power directly linked to upper-class interests, while _____ begin with the premise that an understanding of power cannot be sound without a strong analysis of gender.

   A. Autonomous state theorists; feminist theorists
   B. Power elite theorists; feminist theorists
   C. Pluralist theorists; feminist theorists
   D. none of the above

27. According to the text, the U.S. Senate is _____ percent men.

   A. 59
   B. 22
   C. 76
   D. 87

28. _____ theory sees interest groups compete in a struggle for power, while _____ theory sees power as stemming from the top down.

   A. Power elite; autonomous state
   B. Autonomous state; power elite
   C. Pluralist; power elite
   D. Power elite; pluralist

29. _____ theory sees the power of the state as feeding on itself.

   A. Pluralist
   B. Power elite
   C. Autonomous state
   D. Feminist

30. _____ is a case of the United States expanding its state interests by regulating business not only in the U.S., but also in Mexico and Canada.

    A. The United States Free Trade Agreement
    B. The North American Free Trade Agreement
    C. The Northern Hemisphere Free Trade Agreement
    D. none of the above

31. A _____ is based on the principle of representing all people through the right to vote.

    A. dictatorship
    B. oligarchy
    C. monarchy
    D. democracy

32. In a _____, the authority of the ruler tends to be inherited.

    A. monarchy
    B. oligarchy
    C. democracy
    D. none of the above

33. The _____ includes those state institutions that represent the population, making rules that govern the society.

    A. family
    B. politics
    C. economy
    D. government

34. In most known cases of dictatorship, the ruler is also a man – showing a relationship between dictatorship and:

    A. matriarchy
    B. patriarchy
    C. monarchy
    D. democracy

35. Saddam Hussein ruled through the force of terror and the loyalty of his "_____ party."

    A. *Baath*
    B. *Muslim*
    C. *Jihad*
    D. *Hussein*

Chapter 19

36. Only about _____ of young people actually vote in presidential elections.

    A. a quarter
    B. half
    C. one-third
    D. two-thirds

37. Women are more liberal than men on issues such as:

    A. militarism
    B. gay and lesbian rights
    C. feminism
    D. all of the above

38. According to the text, more women than men voted for _____ in the 2004 presidential election.

    A. George W. Bush
    B. John Kerry
    C. Ralph Nader
    D. none of the above

39. According to the text, 54 percent of those 18 to 29 years old voted for _____ in the 2004 presidential election.

    A. George W. Bush
    B. John Kerry
    C. Ralph Nader
    D. none of the above

40. Which of the following statements accurately characterize the composition of Congress?

    A. Most of the members of Congress are of middle or lower class background
    B. Most of the members of Congress are White men
    C. No members of Congress are millionaires
    D. No members of Congress are lawyers

41. According to the text, _____ percent of Americans say they would vote for a woman for president.

    A. 53
    B. 68
    C. 81
    D. 92

Politics and Government

42. In the United States, the military is the single largest _____, and it accounts for a large portion of the federal budget.

   A. costumer
   B. employer
   C. client
   D. agency

43. According to the text, _____ percent of people in the U.S. say they have quite a lot or a great deal of confidence in the military.

   A. 45
   B. 65
   C. 95
   D. 75

44. The term used to describe the linkage between business and military interests is:

   A. military-industrial complex
   B. the collusion between business and the military
   C. business-defense connection
   D. political graft

45. In _____, the all-male draft was ended and an all-volunteer force was initiated in the U.S. armed services.

   A. 1967
   B. 1973
   C. 1982
   D. 1991

46. According to the text, _____ percent of military personnel are racial minorities.

   A. 67
   B. 19
   C. 58
   D. 34

47. Women are about _____ percent of the forces serving in the Afghanistan and Iraq wars.

   A. 20
   B. 10
   C. 40
   D. 30

48. The first female uniformed personnel appeared in 1901 as:

   A. mechanics
   B. drivers
   C. nurses
   D. soldiers

Chapter 19

49. The "Don't ask, don't tell" policy on dealing with gays and lesbians in the military:

    A. still allows recruiting officers to ask about sexual preference
    B. has, in reality, given gays and lesbians the right to live openly as homosexuals while pursuing their military careers
    C. fails to prevent those who keep their sexual preference a secret from discrimination and/or harassment
    D. still allows the dismissal from the military of gays or lesbians who reveal their sexual preference

50. According to the text, only _____ percent of the public do not think gays should be allowed to serve under any circumstances.

    A. 21
    B. 14
    C. 9
    D. 2

True/False Questions:

T   F   1. The guarantee of life, liberty, and the pursuit of happiness is evenly distributed to each state in the United States.

T   F   2. One of the state's roles in maintaining order is to manage dissent.

T   F   3. The World Trade Organization (WTO) was created in 1964 to monitor and resolve trade disputes.

T   F   4. Legitimate power is achieved through force, often against the will of the people being forced.

T   F   5. Traditional authority is derived from the personal appeal of a leader.

T   F   6. A bureaucracy is a type of formal organization characterized by an authority hierarchy, a clear division of labor, explicit rules, and impersonality.

T   F   7. An interest group is a constituency in society organized to promote its own agenda.

T   F   8. The power elite model was originated in the work of George Herbert Mead.

T   F   9. The autonomous state theory sees the state as taking on a life of its own form and interests.

T   F   10. The feminist theory interprets political power as emerging from the dominance of men over women.

T   F   11. The United States has one of the highest voter turnouts among democratic nations.

Politics and Government

T   F   12.   There is no significant variation in voting patterns by race.

T   F   13.   The gender gap refers to the differences in people's political attitudes and behavior.

T   F   14.   The average cost of a Senate campaign is $3.7 million.

T   F   15.   A record number of women (13) now serve in the U.S. Senate and now over-represent their proportion to the U. S. population.

T   F   16.   Militarism is highly ranked in the value system of the United States.

T   F   17.   Black Americans are the most underrepresented in the military relative to their proportion to the civilian population.

T   F   18.   The military has long been seen as an institution in which race relations are better than among the general public.

T   F   19.   Minority women are overrepresented in the military, relative to their percentage in the general population.

T   F   20.   Military recruiters can interrogate recruits about their sexual preference and orientation.

Short Answer Questions:

1.   Compare and contrast between power and authority.
2.   What are political action committees (PACs)?
3.   Compare and contrast between democracy, monarchy, and dictatorship. Provide examples.
4.   What is the military-industrial complex?

Chapter 19

ANSWERS FOR CHAPTER NINETEEN

Multiple Choice Questions:

1. D (p.510)
2. D (p.510)
3. B (p.511)
4. C (p.512)
5. D (p.512)
6. B (p.512)
7. A (p.512)
8. B (p.513)
9. A (p.513)
10. B (p.513)
11. D (p.513)
12. B (p.514)
13. D (p.514)
14. A (p.514)
15. C (p.515)
16. C (p.515)
17. A (p.515)
18. A (p.516)
19. D (p.516)
20. B (p.516)
21. A (p.517)
22. C (p.517)
23. B (p.519)
24. C (p.519)
25. C (p.519)
26. B (p.519)
27. D (p.519)
28. C (p.519)
29. C (p.519)
30. B (p.519)
31. D (p.520)
32. A (p.520)
33. D (p.520)
34. B (p.521)
35. A (p.521)
36. C (p.521)
37. D (p.523)
38. B (p.523, Table 19.3)
39. B (p.523, Table 19.3)
40. B (p.524)
41. D (p.526)
42. B (p.526)
43. D (p.527)
44. A (p.528)
45. B (p.529)
46. D (p.529)
47. B (p.530)
48. C (p.530)
49. D (pp.530-531)
50. C (p.531)

True/False Questions:

1. F (p.510)
2. T (p.511)
3. F (p.512)
4. F (p.513)
5. F (p.513)
6. T (p.514)
7. T (p.515)
8. F (p.516)
9. T (pp.518-519)
10. T (p.519)
11. F (p.521)
12. F (p.522)
13. T (p.523)
14. T (p.524)
15. F (p.525)
16. T (p.526)
17. F (p.527)
18. T (p.529)
19. T (p.530)
20. F (p.531)

Short Answer Questions:

1. See p.513
2. See p.516
3. See pp.520-521
4. See pp.528-529

CHAPTER TWENTY

HEALTH CARE

Multiple Choice:

1. One insight of the scientific revolution of the mid-1800s was the _____, the idea that many illnesses were caused by microscopic organisms.

    A. magic bullet theory
    B. etiological theory
    C. blood letting
    D. germ theory

2. The American Medical Association (AMA) was founded in:

    A. 1912
    B. 1802
    C. 1898
    D. 1847

3. According to the text, it was in the late _____ that the image of medicine as an upper-class profession took hold.

    A. 1900s
    B. 1800s
    C. 1700s
    D. 1600s

4. Today, about _____ percent of physicians are specialists.

    A. 80
    B. 20
    C. 40
    D. 60

5. The _____ program, begun in 1965, provides medical care in the form of medical insurance covering hospital costs for all individuals who are age 65 or older.

    A. Medisure
    B. Medicaid
    C. Medicare
    D. Eldercare

Chapter 20

6. _____ is the governmental program which provides medical care in the form of health insurance for the poor, those on welfare, and for the disabled.

   A. Medisure
   B. Care for the Needy Individual (CNI)
   C. Medicare
   D. Medicaid

7. The Medicare and _____ programs together are as close as the United States has come to the ideal of universal health insurance.

   A. Medicgap
   B. Medicall
   C. Medicaid
   D. Medicassist

8. From pharmaceutical companies to hospitals, health care is increasingly being delivered in an organizational form where _____ interests dominate.

   A. altruistic
   B. corporate
   C. humanitarian
   D. charitable

9. _____ argues that any institution, group, or organization can be interpreted by looking at its positive and negative functions in society.

   A. Conflict theory
   B. Functionalism
   C. Symbolic interactionism
   D. Feminist theory

10. _____ functions contribute to the harmony and stability of society.

    A. Manifest
    B. Latent
    C. Negative
    D. Positive

11. According to _____, the positive functions of the health care system are the prevention and treatment of disease.

    A. conflict theory
    B. functionalism
    C. symbolic interactionism
    D. feminist theory

12. The _____ perspective primarily examines the institutional system of health care and studies how different institutional forms of health care benefit society as a whole.

    A. conflict
    B. feminist
    C. symbolic interactionist
    D. functionalist

13. According to the _____ theory of health care, the inequality inherent in our capitalist society is responsible for unequal access to medical care.

    A. conflict
    B. functionalist
    C. symbolic interactionist
    D. feminist

14. _____ theorists are interested in how illness and death are distributed across the various groups in society.

    A. Symbolic interaction
    B. Functional
    C. Conflict
    D. Social exchange

15. Which of the following groups is most likely to examine inequality within health care employment patterns?

    A. symbolic interactionists
    B. functionalists
    C. conflict theorists
    D. social exchange theorists

16. _____ theorists are most critical of the corporate control of health care and associate the drive for corporate profits with the rising costs of health care.

    A. Functional
    B. Conflict
    C. Symbolic interaction
    D. Social exchange

17. Which of the following groups is most likely to argue that definitions of illness and wellness are culturally relative?

    A. symbolic interactionists
    B. conflict theorists
    C. functionalists
    D. social exchange theorists

Chapter 20

18. _____ refers to the belief that one is sick even when not.

    A. Multichondria
    B. Hyperchondria
    C. Hypochondria
    D. none of the above

19. According to _____ theorists, illness is, in part, socially constructed.

    A. conflict
    B. functionalist
    C. symbolic interaction
    D. social exchange

20. According to _____ theorists, policy should decrease the negative functions of the health care system.

    A. symbolic interaction
    B. conflict
    C. feminist
    D. functionalist

21. The fundamental problem uncovered by _____ theory is that patients are patronized and infantilized.

    A. conflict
    B. symbolic interaction
    C. functional
    D. social exchange

22. The fundamental problem uncovered by _____ theory is the excessive bureaucratization of the health care system.

    A. functional
    B. symbolic interaction
    C. conflict
    D. social exchange

23. _____ also studies how the interaction between a physician and patient can be "managed."

    A. Symbolic interaction
    B. Conflict theory
    C. Functionalism
    D. Structural theory

24. On average, women tend to receive _____ quality of health care than men, even though they tend to utilize the health care system _____.

   A. very high; more
   B. a higher; less
   C. a lesser; more
   D. very low; less

25. White women can expect to live longer than White men, to _____ years of age.

   A. 78.9
   B. 75.5
   C. 80.2
   D. 86.4

26. Which of the following groups has the highest life expectancy?

   A. White men
   B. White women
   C. African American women
   D. African American men

27. Which of the following groups has the lowest life expectancy?

   A. White men
   B. White women
   C. African American men
   D. African American women

28. _____ has the federal responsibility for the health of the Native American population.

   A. Indian Health Services
   B. Department of Health and Human Services
   C. Institutes of Health
   D. Centers for Disease Control

29. The lower the social class status of the person or family:

   A. the less likely they are to qualify for assistance
   B. the less access they have to adequate health care
   C. the more access they have to adequate health care
   D. the more likely they are to be in great health

30. According to the text, _____ percent of the population have no health insurance at all.

   A. 14.5
   B. 18.3
   C. 24.7
   D. 31.2

Chapter 20

31. _____ have a higher likelihood of developing chronic disease than _____, although _____ are more likely to be disabled by disease.

    A. The young; the elderly; the young
    B. Men; women; men
    C. Women; men; men
    D. none of the above

32. According to the text, under the age of _____, men are more likely to be overweight than women.

    A. 55
    B. 35
    C. 45
    D. 65

33. As income inequality within a nation increases, so do:

    A. rates of infant mortality
    B. lack of health care insurance
    C. more low-birth-weight babies
    D. all of the above

34. Which of the following countries has a life expectancy of barely 45 years of age?

    A. Somalia
    B. Laos
    C. Chad
    D. all of the above

35. An eating disorder characterized by compulsive dieting is:

    A. anorexia nervosa
    B. carnivitis
    C. bulimia
    D. the Ally McBeal syndrome

36. An eating disorder characterized by alternate binge eating and then purging or induced vomiting in order to lose weight is:

    A. bulimia
    B. compulsive overeating
    C. anorexia nervosa
    D. carnivitis

37. Obesity has traditionally been considered a matter of individual habits, but in _____ the officials in the federal Medicare program changed their policy to include obesity as a disease.

    A. 2000
    B. 1998
    C. 2004
    D. 2002

38. Many athletes, both professional and amateur, have been goaded by athletic dreams to use powerful hormones to stimulate the growth of muscle. These chemicals are:

   A. amphetamines
   B. herbal remedies
   C. anabolic steroids
   D. nutritional supplements

39. Stigmas:

   A. are a part of HIV-positive status
   B. occur when an individual is socially de-valued because of some malady, illness, misfortune, or similar attribute
   C. occur for every medical attribute
   D. both A and B

40. According to the text, approximately _____ sexually transmitted diseases have been medically diagnosed.

   A. 10
   B. 25
   C. 50
   D. 40

41. The number of AIDS cases in 2003 was highest among:

   A. White females
   B. Hispanic males
   C. African males
   D. Native American females

42. The AIDS disease is transmitted through the exchange of bodily fluids, particularly:

   A. blood
   B. semen
   C. tears
   D. both A and B

43. According to the text, how the mentally ill are perceived by the medical profession and the public depends in significant part on:

   A. the physical symptoms associated with the illness
   B. the label that is attached to various behaviors
   C. the ability to treat the illness
   D. none of the above

Chapter 20

44. Treatment of a patient that might prolong life or a coma but offer no hope of recovery is called:

   A. heroic treatment
   B. futile treatment
   C. terminal treatment
   D. braindead treatment

45. The United States tops the list of all countries in _____ expenditures for health care.

   A. local government
   B. per-person
   C. preventative
   D. none of the above

46. Currently, the cost of medical care in the U.S. is more than _____ percent of our gross national product.

   A. 14
   B. 8
   C. 15
   D. 17

47. Sweden and _____ spend roughly half as much per capita on medical care as the United States.

   A. Britain
   B. Mexico
   C. Egypt
   D. France

48. Overall, _____ million people in the United States are without health insurance.

   A. 73
   B. 23
   C. 58
   D. 41

49. Which of the following statements is true?

   A. currently, the cost of medical care in the United States is more than 31 percent of our gross national product
   B. the cost of healthcare has, at times, risen at twice the inflation rate
   C. while expensive, the U.S. health-care system, is undisputedly the best in the world
   D. all of the above

50. Which of the following statements is (are) true about medical malpractice?

   A. the number of patients who sue their physicians is decreasing
   B. doctors often practice defensive medicine to prevent against malpractice suits
   C. lawyers despise and run away from malpractice suits
   D. none of the above

Health Care

True/False Questions:

T F 1. The year 1847 saw the founding of the American Medical Association (AMA), and is the most powerful organization in U.S. health care.

T F 2. The Medicare program began as an aftermath of the Civil War to care for veterans.

T F 3. National Health Care is a governmental program that provides medical care in the form of health insurance for the poor, those on welfare, and for the disabled.

T F 4. Conflict theorist hold that medical practitioners frequently subject patients to infantilization.

T F 5. U.S. citizens are unhealthy in relation to the rest of the world.

T F 6. Anorexia nervosa is an eating disorder characterized by compulsive dieting.

T F 7. Bulimia is an eating disorder characterized by alternate binge eating then purging to lose weight.

T F 8. Men have been exempt from the pressure of the social values placed on physical appearance.

T F 9. Some 450,000 people die each year as a direct result of smoking.

T F 10. AIDS is the term for a category of disorders that result from a breakdown of the body's immune system.

T F 11. AIDS is the number one sexually transmitted disease (STD) in the United States.

T F 12. The U.S. Supreme Court restricted the definition of disability to exclude disabilities that can be corrected with devices such as eyeglasses or with medication.

T F 13. Terminally ill patients and a close family member have the right to refuse what is called "heroic" treatment of the patient that might prolong life.

T F 14. Negative euthanasia is often called "mercy killing."

T F 15. The U.S. tops the list of all countries in per-person expenditures for health care, and yet our health care system is not the best in the world.

T F 16. There are over 41 million people in the U.S. without any health care insurance.

T F 17. A living will is a statement made by the patient while in possession of all mental faculties as to whether or not heroic treatment ought to be given in the case of severe incapacity.

T F 18. HMOs are experiencing decline due to widespread fraud and abuse in the medical community.

Chapter 20

T F 19. Holistic medicine emphasizes the person's entire mental, physical, social, well-being.

T F 20. Practitioners of holistic medicine reject the remedies of traditional medicine such as surgery and drugs.

Short Answer Questions:

1. Compare and contrast between Medicare and Medicaid.
2. Compare and contrast between anorexia nervosa and bulimia.
3. Describe the medical model. Provide examples.
4. What is defensive medicine? Provide examples.

Health Care

ANSWERS FOR CHAPTER TWENTY

Multiple Choice Questions:

1. D (p.536)
2. D (p.537)
3. B (p.537)
4. A (p.537)
5. C (p.538)
6. D (p.538)
7. C (p.538)
8. B (p.539)
9. B (p.539)
10. D (p.539)
11. B (p.539)
12. D (p.539)
13. A (p.539)
14. C (p.540)
15. C (p.540)
16. B (p.540)
17. A (p.540)
18. C (p.540)
19. C (p.540)
20. D (p.540)
21. B (p.540, Table 20.1)
22. C (p.540, Table 20.1)
23. A (p.541)
24. C (p.541)
25. C (p.541)
26. B (p.541)
27. C (p.541)
28. A (p.542)
29. B (p.542)
30. A (p.543)
31. C (p.543)
32. B (p.543)
33. D (p.544)
34. D (p.544)
35. A (p.545)
36. A (p.545)
37. C (p.547)
38. C (p.547)
39. D (p.548)
40. C (p.548)
41. C (p.549)
42. C (p.549)
43. B (p.551)
44. A (p.552)
45. B (p.553)
46. A (p.553)
47. D (p.553)
48. D (p.553)
49. B (p.553)
50. B (p.554)

True/False Questions:

1. T (p.537)
2. F (p.538)
3. F (p.538)
4. F (p.540)
5. F (p.541)
6. T (p.545)
7. T (p.545)
8. F (p.546)
9. T (p.547)
10. T (p.548)
11. F (p.548)
12. T (p.550)
13. T (p.552)
14. F (p.552)
15. T (p.553)
16. T (p.553)
17. T (p.553)
18. F (p.555)
19. T (p.555)
20. F (p.555)

Short Answer Questions:

1. See pp.538-539
2. See pp.545-546
3. See p.547
4. See p.554

Chapter 21

CHAPTER TWENTY-ONE

POPULATION, URBANIZATION, AND THE ENVIRONMENT

Multiple Choice:

1. The Baby Boom is the name given to that crop of _____ babies currently representing nearly one-third of all the people in the United States.

    A. 95 million
    B. 100 million
    C. 50 million
    D. 75 million

2. Among those most likely to be missed by the census are:

    A. the homeless
    B. immigrants
    C. minorities living in ghettos
    D. all of the above

3. The total number of people in a society at any given moment is determined by three variables: births, deaths, and:

    A. emigrations
    B. migrations
    C. immigrations
    D. none of the above

4. According to the text, the population of the entire world is increasing at a rate of about _____ people per day.

    A. 780,000
    B. 270,000
    C. 120,000
    D. 560,000

5. Current demographic estimates are that a majority of women could have a maximum average of nearly _____ children.

    A. twenty-five
    B. fifteen
    C. ten
    D. twenty

Population, Urbanization, and the Environment

6. The number of babies born each year for every one thousand members of the population is the:

    A. crude birth rate
    B. fertility rate
    C. infant percentile rate
    D. none of the above

7. According to the text, 16.5 births per 1000 people, is the current birthrate for:

    A. White
    B. African Americans
    C. the United States
    D. Latinos

8. Which of the following groups has the highest birthrate (per 1000)?

    A. African Americans
    B. Latinos
    C. Native Americans
    D. Asian Americans

9. One of the countries with the lowest birthrate in the world is:

    A. Italy
    B. United States
    C. Canada
    D. Mexico

10. Given the current birthrates, the population of African Americans will double by the year:

    A. 2030
    B. 2025
    C. 2050
    D. 2015

11. Life expectancy:

    A. is defined as the average number of years that a group can expect to live
    B. in the United States is over 75 years today
    C. varies with gender, race-ethnicity, and social class
    D. all of the above

12. According to the text, Afghanistan has an infant mortality rate of:

    A. 200 infant deaths per 1000 live births
    B. 150 infant deaths per 1000 live births
    C. 100 infant deaths per 1000 live births
    D. 50 infant deaths per 1000 live births

Chapter 21

13. Which of the following countries has a life expectancy higher than the United States?

    A. Australia
    B. Spain
    C. Canada
    D. all of the above

14. _____ babies are almost *thirty* times more likely to contract AIDS than White babies.

    A. Asian American
    B. Hispanic American
    C. African American
    D. Native American

15. Among Hispanics, migration patterns have been traditionally linked to the:

    A. agriculture industry
    B. auto industry
    C. construction industry
    D. oil industry

16. In the United States, there are _____ males for every 100 females.

    A. 105
    B. 87
    C. 78
    D. 94

17. In the United States, more and more people are entering the sixty-five-and-over age bracket. This trend is knows as:

    A. the graying of America
    B. the retirement of America
    C. the slowing of America
    D. none of the above

18. Which of the following statements is (are) true about birth cohorts?

    A. the baby boom generation currently makes up one-half of the population
    B. cohorts can stay the same, increase, or decrease in size
    C. birth cohorts consist of all people born within a given period
    D. both B and C

19. For encouraging parents to communicate with their children and forego physical punishments when possible in the late 1950s and 1960s, _____ has often been blamed at one time or another for nearly every social problem at the time.

    A. Dr. Phil
    B. Dr. Spock
    C. Dr. Laura
    D. Dr. Malthus

20. Malthus considered _____ to be a preventative check on population growth.

    A. war
    B. famine
    C. sexual abstinence
    D. disease

21. The population theory which argued that a population tends to grow faster than the subsistence needed to sustain it is:

    A. Malthusian theory
    B. the "population bomb" theory
    C. the demographic transition theory
    D. the family planning theory

22. Humans, like other animals, can survive and reproduce only when they have access to the means of _____, meaning the necessities of life.

    A. production
    B. saturation
    C. subsistence
    D. cessation

23. According to the text, the development and widespread use of _____ in many countries have kept the birthrate at a level lower than Malthus would have thought possible.

    A. contraconceptives
    B. contraceptives
    C. contraception
    D. none of the above

24. During Malthus's times, there were _____ children per family.

    A. 10
    B. 4
    C. 7
    D. 2

Chapter 21

25. Demographic transition theory:

   A. proposes that countries pass through a consistent sequence of population patterns linked to the degree of development in the society
   B. predicts that the population level will continually rise
   C. theorizes that populations fluctuate in a cyclical fashion
   D. both A and C

26. In the demographic transition theory, which stage is characterized by an overall increase in population growth?

   A. Stage 1
   B. Stage 4
   C. Stage 3
   D. Stage 2

27. _____ theory was used to counter the predictions of demographic transition theory.

   A. Social exchange
   B. Autonomous choice
   C. Rational choice
   D. none of the above

28. The state in which the combined birth and death rate of a population simply sustains the population at a steady level is:

   A. the equal population level
   B. the limited population growth amount
   C. the population replacement level
   D. the Stage 2 population level

29. ZPG is dedicated to reaching the _____, a state in which the combined birthrate and death rate of a population simply sustains the population at a steady level.

   A. equilibrium replacement level
   B. population replacement level
   C. cohort replacement level
   D. none of the above

30. By _____, the United States reached the replacement level of reproduction with an overall average number of 2.1 children per family.

   A. 1970
   B. 1980
   C. 1960
   D. 1940

Population, Urbanization, and the Environment

31. _____ are taught that it is acceptable to use natural means of birth control but are forbidden to use contraceptive devices.

    A. Protestants
    B. Catholics
    C. Jewish
    D. none of the above

32. Studies of family planning indicate that:

    A. creating a new image of the ideal family is a central concern of those involved in the family planning movement
    B. the majority of Catholics still limit their birth control efforts to coitus interruptus, the rhythm method, and abstinence
    C. there is no firm data to confirm the demographic theory's contention that as countries in general become more economically developed, their birthrates and average family size drop
    D. all of the above are true

33. Birthrate and family size are known to be related to the overall level of:

    A. economic development of a country
    B. political development of a country
    C. religious development of a country
    D. none of the above

34. Governmental programs that advocate contraception can be successful only if couples:

    A. abstain from sexual intercourse
    B. recognize the government as legitimate
    C. choose to have smaller families
    D. targeted are legally married

35. The 1924 National Origins Quota Law discouraged immigration from the following countries except:

    A. Turkey
    B. Greece
    C. Poland
    D. Norway

36. According to Gans, today's urban underclass would encompass what he called:

    A. the *poor cosmopolites*
    B. the poor villagers
    C. the *downward-bound*
    D. the *trapped*

Chapter 21

37. The practice of _____ by banks further intensifies residential segregation.

   A. steering
   B. redlining
   C. minority transfer
   D. redlighting

38. The subfield of sociology that examines the social structure and cultural aspects of life in the city in comparison to rural and suburban centers is:

   A. suburban sociology
   B. rural sociology
   C. street-corner sociology
   D. urban sociology

39. The 1924 National Origins Quota Law encouraged immigration from the following countries except:

   A. England
   B. France
   C. Germany
   D. Italy

40. Today's urban underclass would encompass the group that Gans called the _____ of late 1950's Boston.

   A. cosmopolites
   B. ethnic villagers
   C. forgotten
   D. trapped

41. _____ is determined by the number of people per unit of area, usually per square mile.

   A. Population paucity
   B. Population density
   C. Population ecology
   D. Population divide

42. Which of the following is an example of a human ecosystem?

   A. Tokyo
   B. Tampa
   C. Mexico City
   D. all of the above

43. Two fundamental and closely related problems confront our present ecosystems: the destruction/exhaustion of natural resources and:

    A. pollution
    B. global warming
    C. overpopulation
    D. none of the above

44. If one element of an ecosystem is disturbed:

    A. nothing happens
    B. the entire system is affected
    C. only parts of the system are affected
    D. none of the above

45. According to the text, if a growing population is a problem of the developing world, then _____ a problem of the industrialized world.

    A. the expansion of arable land is
    B. shrinking resources are
    C. negative population growth is
    D. vast sources of water are

46. The leading air and water polluters are the:

    A. the United States
    B. Japan
    C. Poland
    D. all of the above

47. The EPA has estimated that hazardous and cancer-causing pollutants released into the air by industry are responsible for approximately _____ or more deaths a year.

    A. 20,000
    B. 2000
    C. 200,000
    D. 200

48. _____ occurs when industries take water from the rivers and lakes to use in their production processes and return the water heated up and polluted, causing differences in temperature that can alter aquatic habitats and kill aquatic life.

    A. Chemical pollution
    B. Industrial pollution
    C. Thermal pollution
    D. Air pollution

Chapter 21

49. Studies of the response of men and women to environmental issues indicate that:

    A. men are more concerned than women about global warming
    B. women consistently showed more concern than men for environmental issues, and also perceived themselves to be considerably more at risk from environmental hazards than men
    C. men were more apt than women to feel that waste sites posed dangers to trees, fish, and other wildlife
    D. men and women are equally concerned about these issues

50. The Office of Population Research, founded in 1936 at _____ University, is the oldest population research center in the country.

    A. Harvard
    B. Florida State
    C. Princeton
    D. Northwestern

True/False Questions:

T F 1. Vital statistics include information such as birth date, age, weight, and Zodiac sign.

T F 2. Emigration refers to the departure of a people from a society.

T F 3. The fertility of a population has to do with the potential number of children in a population that could be born.

T F 4. The United States has the highest life expectancy and the lowest infant mortality rate in industrialized countries.

T F 5. Migration can not occur within the boundaries of a country.

T F 6. Sex and age data are often combined in a graphical format called an age-sex pyramid.

T F 7. A birth cohort consists of all the persons born within a given period.

T F 8. Malthusian theory is the most positive and upbeat of all global population theories.

T F 9. There is no connection between the relationship of economic development and family size in demographic transition theory.

T F 10. Ehrlich argues that the availability of clean air and water are the critical factors in the growth and health of populations.

T F 11. Equilibrium level refers to a balanced level between birth and death rates, and relates to Zero Population Growth (ZPG).

Population, Urbanization, and the Environment

T   F   12.   The rural-urban continuum refers to the growth and development of cities and population density.

T   F   13.   The largest number of immigrants to the U.S. recently have been from Mexico, Philippines, China, and countries of the former Soviet Union.

T   F   14.   Global Organization for a Better Environment (GLOBE) is an international organization that is attempting to stop all forms of pollution, particularly water pollution.

T   F   15.   Environmental racism refers to radicals who are out creating toxic genocide.

T   F   16.   Among the largest commercial hazardous waste landfill in the nation is located in Emelle, Alabama.

T   F   17.   The environmental movement has not affected environmental policy of the U.S. government.

T   F   18.   The field of ecological demography combines the studies of demography and ecology.

T   F   19.   Ecological globalization focuses on the worldwide dispersion of problems and issues involving the relationships between humans and the physical and social global environment.

T   F   20.   Environmental concerns are not found on international political agendas.

Short Answer Questions:

1. Briefly describe Malthusian theory.
2. List and discuss the three stages of demographic transition theory.
3. Define environmental racism. Provide examples.
4. Define ecological demography.

Chapter 21

ANSWERS FOR CHAPTER TWENTY-ONE

Multiple Choice Questions:

1. D (p.561)
2. D (p.562)
3. B (p.562)
4. B (p.563)
5. D (p.563)
6. A (p.563)
7. C (p.563)
8. B (p.563)
9. A (p.563)
10. B (p.564)
11. D (p.564)
12. B (p.564)
13. D (p.564, Table 21.1)
14. C (p.565)
15. A (p.565)
16. D (p.566)
17. A (p.566)
18. C (p.567)
20. B (p.567)
19. C (p.568)
21. A (p.568)
22. C (p.568)
23. B (p.568)
24. C (p.569)
25. A (p.569)
26. D (p.569)
27. C (p.570)
28. C (p.570)
29. B (p.570)
30. B (p.571)
31. B (p.571)
32. A (p.572)
33. A (p.572)
34. C (p.572)
35. D (p.573)
36. D (p.573)
37. B (p.573)
38. D (p.573)
39. D (p.573)
40. D (p.573)
41. B (p.574)
42. D (p.574)
43. C (p.574)
44. B (p.574)
45. B (p.575)
46. D (p.576)
47. B (p.576)
48. C (p.578)
49. B (pp.581-582)
50. C (p.582)

True/False Questions:

1. F (p.562)
2. T (p.563)
3. F (p.563)
4. F (p.564)
5. F (p.565)
6. T (p.566)
7. T (p.567)
8. F (p.568)
11. F (p.569)
9. T (p.570)
10. T (pp.570-571)
12. F (p.573)
13. T (p.574)
14. T (p.579)
15. F (p.579)
16. T (p.580)
17. F (p.582)
18. T (pp.584-585)
19. T (p.585)
20. F (p.585)

Short Answer Questions:

1. See pp.568-569
2. See pp.569-570
3. See pp.579-581
4. See pp.584-585

CHAPTER TWENTY-TWO

COLLECTIVE BEHAVIOR AND SOCIAL MOVEMENTS

Multiple Choice:

1. Collective behavior occurs when something _____ happens and people respond by establishing new behavioral norms.

    A. ordinary
    B. common
    C. mundane
    D. out of the ordinary

2. Collective behavior:

    A. is behavior that occurs when usual conventions are suspended and people collectively establish new norms of behavior in response to an emerging situation
    B. can occur in response to the strong belief that an event will occur
    C. includes crowds, riots, and panics
    D. all of the above

3. Collective behavior refers to:

    A. highly emotional, even volatile situations
    B. the proliferation of rumors
    C. unusual or unexpected situations
    D. all the above

4. The _____ movement includes a wide array of communities that have organized to protest the dumping and pollution that imperils their neighborhoods.

    A. environmental peace
    B. environmental justice
    C. environmental equity
    D. environmental integrity

5. From a sociological point of view, _____ are the information transmitted by participants in collective behavior as they try to make sense out of an ambiguous situation.

    A. propaganda
    B. chichat
    C. rumors
    D. facts

Chapter 22

6. According to the text, crowds usually have a sense of:

    A. *urgency*
    B. *holding back*
    C. *triviality*
    D. *irrelevance*

7. The people closest to the crowd's center of interest are the _____ of the crowd and show the greatest focus on the object of interest.

    A. *periphery*
    B. *fringe*
    C. *core*
    D. *edge*

8. Crowds may generate _____ *crowds*, groups that may be physically present or that may be as remote as a mass media audience.

    A. *secondary*
    B. *bystander*
    C. *sub-*
    D. *derivative*

9. Which of the following is an example of an expressive crowd?

    A. a victory celebration
    B. a religious revival
    C. a funeral procession
    D. all of the above

10. Early social theorists such as Gustave LeBon described crowds as having mental unity or a:

    A. "shared mind"
    B. "group mind"
    C. "collective mind"
    D. none of the above

11. _____ postulates that, when people are faced with an unusual situation, they create meanings that define and direct the situation.

    A. Evolving consciousness theory
    B. Involving norm theory
    C. Collective consciousness theory
    D. Emergent norm theory

12. Expressive crowds are groups whose primary function is:

    A. the release or expression of emotion
    B. the overthrow of the government
    C. to save certain animals from going extinct
    D. social bonding

13. According to the text, _____ main factors characterize panic-producing situations.

    A. two
    B. three
    C. four
    D. five

14. Behavior that occurs when the people in a group suddenly become concerned for their safety, and seemingly spontaneous, disorganized behavior results is called:

    A. panic
    B. social trauma
    C. social pandemonium
    D. collective surge

15. Which of the following statements is (are) true about panic?

    A. There is easy front-to-rear communication in panic situations
    B. In panic, there is no social structure
    C. Panic occurs when the people in a group suddenly become concerned about their safety
    D. all of the above

16. In the sinking of the Titanic in 1912, of the people traveling in third class, ____ percent of the women died.

    A. 75
    B. 85
    C. 45
    D. 25

17. In the sinking of the Titanic in 1912, of the people traveling in third class, ____ percent of the children died.

    A. 85
    B. 70
    C. 60
    D. 45

18. In the sinking of the Titanic in 1912, of the 2200 people onboard, _____ escaped in lifeboats.

    A. 1000
    B. 900
    C. 800
    D. 600

Chapter 22

19. Three main factors characterize panic-producing situations: a perceived threat, possible entrapment, and:

   A. fainting spells
   B. uncontrollable fatigue
   C. a failure of front-to-rear communication
   D. inability to speak

20. Sociologists see *riots* as a multitude of small _____ spread over a particular geographic area.

   A. hysterical actions
   B. panic attacks
   C. crowd actions
   D. none of the above

21. Most riot activity occurs in the _____, suggesting that riot behavior is linked to other social routines.

   A. middle of the afternoon
   B. morning and early afternoon
   C. evening and late at night
   D. none of the above

22. Explanations of riots that focus on individual attitudes and states of mind fall into the category of:

   A. rational exchange theory
   B. competition theory
   C. predisposition theory
   D. convergence theory

23. Riots are more likely to occur in cities:

   A. where grievances of rioting group have not been addressed
   B. with economic deprivation of racial-ethnic minorities
   C. with rapid influx of new populations
   D. all of the above

24. The _____ theory argues that people riot when they have to compete for limited resources.

   A. relative deprivation
   B. emergent norm
   C. competition
   D. personal needs

25. Which of the following statements are possible explanations for what brings riots to an end?

    A. the actions of social control agents end the violence
    B. the goals of the protest group have been ignored
    C. the political situation remains stable
    D. all of the above

26. _____ are forms of collective behavior wherein many people over a relatively broad spectrum engage in similar behavior and have a shared definition of their behavior as needed to bring social change or identify their place in society.

    A. Collective consciousness
    B. Collective preoccupations
    C. Collective mood
    D. Collective mindset

27. _____ represent change that has less consequential impact than other social change.

    A. Fads
    B. Fashions
    C. Styles
    D. none of the above

28. Which of the following is an example of a fad?

    A. scooters
    B. raves
    C. hula hoops
    D. all of the above

29. Which of the following is not an example of a fad?

    A. streaking
    B. yo-yos
    C. the Macarena
    D. mini skirt

30. _____ are similar to fads except that they tend to represent more intense involvement for participants.

    A. Drifts
    B. Modes
    C. Crazes
    D. none of the above

Chapter 22

31. In the peaking period of a fad:

   A. no one knows about the fad or activity
   B. the new item is defined as a fad, and people enthusiastically accept it
   C. the fad fades
   D. symptoms of an illness are spread throughout the group

32. Which of the following is an example of conspicuous consumption?

   A. a Lincoln Navigator
   B. a Mercedes Benz
   C. a BMW
   D. all of the above

33. Which of the following statements about scapegoating is (are) true?

   A. Scapegoating often occurs in societies at war
   B. The dominant group is often the victim of scapegoating
   C. Scapegoating occurs when a group collectively identifies another group as a threat to the perceived social order and incorrectly blames the other group for problems that they have not actually caused
   D. both A and B

34. Social movements:

   A. involve a society physically moving to a more pleasant location
   B. involve only small numbers of people
   C. are organized social groups that act with some continuity and coordination to promote or resist change in society or other social units
   D. all of the above

35. Which of the following is an example of a personal transformation movement?

   A. the New Age movement
   B. the environmental movement
   C. the Civil Rights movement
   D. the gay and lesbian movement

36. Which of the following is an example of a reactionary movement?

   A. the New Right
   B. Greenpeace
   C. the Sierra Club
   D. the feminist movement

37. The _____ in particular has become increasingly important for the mobilization of social movements.

    A. radio
    B. telephone
    C. Internet
    D. none of the above

38. _____ is the process by which social movements and their leaders secure people and resources for a movement.

    A. Financial planning
    B. Resource procurement
    C. Mobilization
    D. Finding donors

39. Social movements choose their political and social strategies based on a number of variables including:

    A. the resources available to them
    B. the constraints on their actions
    C. the organizational structure of the movement
    D. all of the above

40. Women of the Ku Klux Klan draw on their familial and community ties to form "_____," groups that use social networks to spread rumor or slander.

    A. filial squads
    B. familial liaisons
    C. mother-in-groups
    D. poison squads

41. Which of the following is part of the White supremacist movement?

    A. the Ku Klux Klan
    B. the neo-Nazis
    C. Christian Identity
    D. all of the above

42. _____ theory emphasizes the vulnerability of the political system to social protest.

    A. Autonomous Movement
    B. New Social Movement
    C. Resource Mobilization
    D. Political Process

Chapter 22

43. _____ theory emphasizes the interconnection between social structural and cultural perspectives.

   A. Autonomous Movement
   B. Political Process
   C. Resource Mobilization
   D. New Social Movement

44. _____ theory emphasizes the linkages among groups within a movement.

   A. Autonomous Movement
   B. Resource Mobilization
   C. New Social Movement
   D. Political Process

45. Which of the following is a resource that can be used to organize a social movement?

   A. money
   B. communication technology
   C. legal knowledge
   D. all of the above

46. _____ theory is used to explain the development of the civil rights movement, which relied heavily on Black churches and colleges for resources.

   A. Autonomous Movement
   B. New Social Movement
   C. Structural Strain
   D. Resource Mobilization

47. Which of the following theories of social movements is an explanation of how social movements develop that focuses on how movements gain momentum by successfully gathering resources, competing with other movements, and mobilizing the resources available to them?

   A. the resource mobilization theory
   B. the new social movement theory
   C. the political process theory
   D. functionalist theory

48. Sociologist Neil Smelser developed _____ theory.

   A. autonomous movement
   B. structural strain
   C. resource mobilization
   D. political process

Collective Behavior and Social Movements

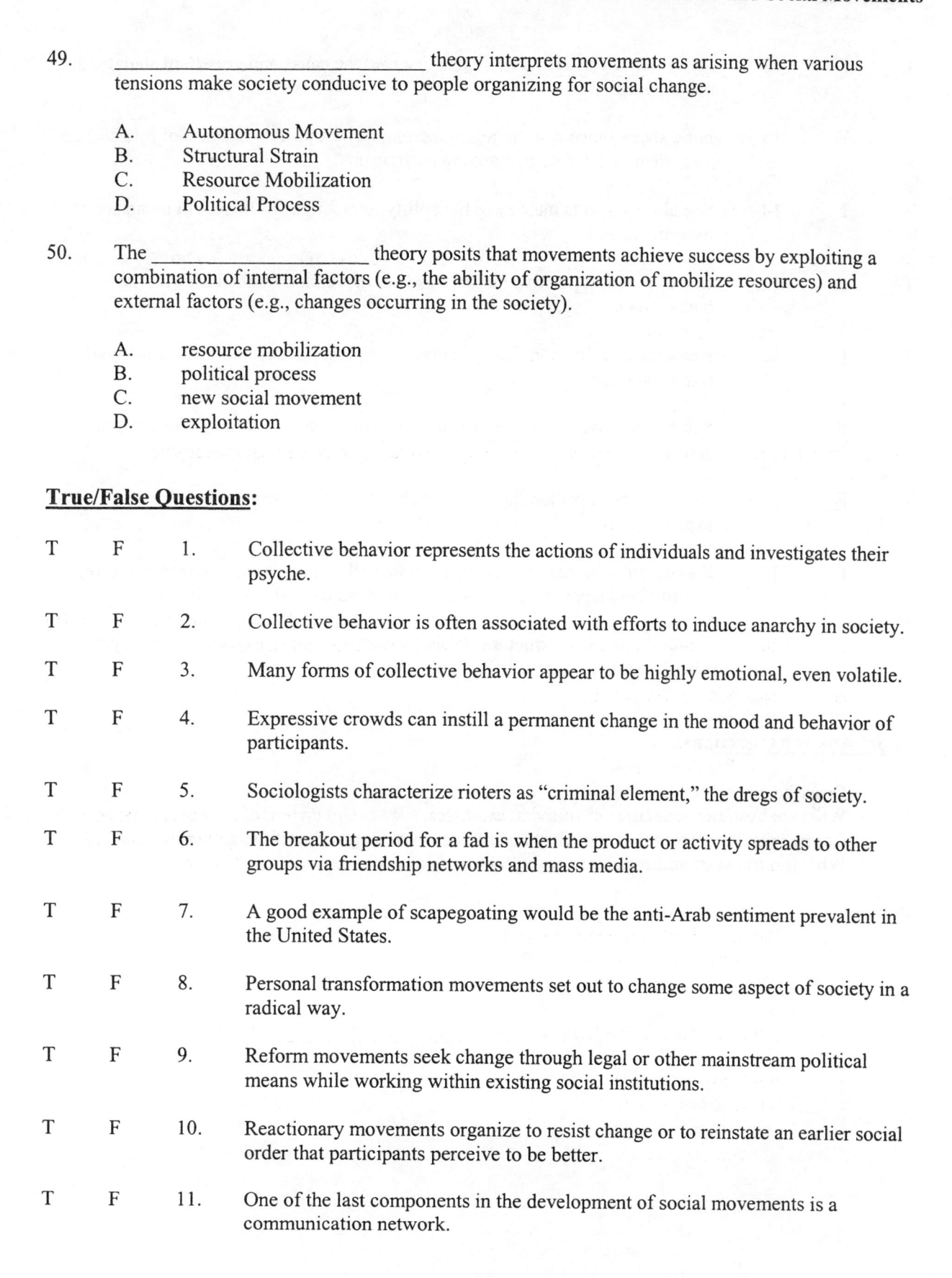

49. _____ theory interprets movements as arising when various tensions make society conducive to people organizing for social change.

    A. Autonomous Movement
    B. Structural Strain
    C. Resource Mobilization
    D. Political Process

50. The _____ theory posits that movements achieve success by exploiting a combination of internal factors (e.g., the ability of organization of mobilize resources) and external factors (e.g., changes occurring in the society).

    A. resource mobilization
    B. political process
    C. new social movement
    D. exploitation

True/False Questions:

T F 1. Collective behavior represents the actions of individuals and investigates their psyche.

T F 2. Collective behavior is often associated with efforts to induce anarchy in society.

T F 3. Many forms of collective behavior appear to be highly emotional, even volatile.

T F 4. Expressive crowds can instill a permanent change in the mood and behavior of participants.

T F 5. Sociologists characterize rioters as "criminal element," the dregs of society.

T F 6. The breakout period for a fad is when the product or activity spreads to other groups via friendship networks and mass media.

T F 7. A good example of scapegoating would be the anti-Arab sentiment prevalent in the United States.

T F 8. Personal transformation movements set out to change some aspect of society in a radical way.

T F 9. Reform movements seek change through legal or other mainstream political means while working within existing social institutions.

T F 10. Reactionary movements organize to resist change or to reinstate an earlier social order that participants perceive to be better.

T F 11. One of the last components in the development of social movements is a communication network.

Chapter 22

T F 12. As a rule, celebrities usually have deterred the cause and growth of social movements.

T F 13. For a social movement to begin, there must be a perceived sense of injustice and a strong desire for change among participants.

T F 14. Social movements must have the ability to mobilize, or there will be no social movement.

T F 15. One of the most positive factors in social movements is the lack of large bureaucracies.

T F 16. Resource mobilization theory is used to explain the development of the civil rights movements.

T F 17. Sociologists use the concept of 'framing' to explain the process for collective action.

T F 18. Resource mobilization theory and political process theory are social structural explanations.

T F 19. Frames are schemes of interpretation that allow people in groups to perceive, identify, and label events that can become the basis for collective action.

T F 20. Cultural and social structural explanations of social movements come together in what is now called new millennium theory.

Short Answer Questions:

1. Describe emergent norm theory.
2. What are hysterical contagions? Provide examples.
3. Compare and contract between reform and radical movements. Provide examples.
4. What is a transnational social movement? Provide examples.

Collective Behavior and Social Movements

ANSWERS FOR CHAPTER TWENTY-TWO

Multiple Choice Questions:

1. D (p.589)
2. D (pp.590-592)
3. D (pp.590-592)
4. B (p.591)
5. C (p.592)
6. A (p.592)
7. C (p.593)
8. B (p.593)
9. D (p.593)
10. B (p.593)
11. D (p.593)
12. A (p.593)
13. B (p.594)
14. A (p.594)
15. C (p.594)
16. C (p.594)
17. B (p.594)
18. D (p.594)
19. C (pp.594-595)
20. C (p.595)
21. C (p.595)
22. D (p.596)
23. D (p.596)
24. C (p.596)
25. A (pp.596-597)
26. B (p.597)
27. A (p.597)
28. D (p.597)
29. D (p.597)
30. C (p.598)
31. B (p.598)
32. D (p.599)
33. C (pp.599-600)
34. C (p.600)
35. A (p.601)
36. A (p.603)
37. C (p.604)
38. C (p.605)
39. D (pp.607-608)
40. D (p.608)
41. D (p.608)
42. D (p.609, Table 22.1)
43. D (p.609, Table 22.1)
44. B (p.609, Table 22.1)
45. D (p.609)
46. D (p.609)
47. A (p.609)
48. B (p.610)
49. B (p.610)
50. B (p.610)

True/False Questions:

1. F (pp.590-592)
2. F (p.590-592)
3. T (p.591)
4. T (p.593)
5. F (p.595)
6. T (p.597)
7. T (pp.599-600)
8. F (p.601)
9. T (p.602)
10. T (p.603)
11. F (p.604)
12. F (p.605)
13. T (p.605)
14. T (pp.605-606)
15. F (p.607)
16. T (p.609)
17. T (p.610)
18. T (p.610)
19. T (p.610)
20. F (p.611)

Short Answer Questions:

1. See p.593
2. See p.599
3. See pp.602-603
4. See p.612

CHAPTER TWENTY-THREE

SOCIAL CHANGE IN GLOBAL PERSPECTIVE

Multiple Choice:

1. The alteration of social interactions, institutions, stratification systems, and elements of culture is:

    A. cultural lag
    B. social variation
    C. innovation
    D. social change

2. Subtle alterations in the day-to-day interaction between people are called:

    A. secondary changes
    B. minichanges
    C. microchanges
    D. indirect changes

3. A fad "catching on" is an example of a:

    A. indirect change
    B. minichange
    C. secondary change
    D. microchange

4. The symptoms of _____ can be seen in the uneven dissemination of computer capability.

    A. cultural delay
    B. cultural lag
    C. cultural holdup
    D. none of the above

5. The principle of "culture lag" is a term originally coined by:

    A. Ferdinand Tönnies
    B. William Ogburn
    C. Immanuel Wallerstein
    D. William McCord

Social Change in Global Perspective

6. Social change generally has a number of shared characteristics, including which of the following?

    A. the onset and consequences of social change cannot be accurately predicted by sociologists
    B. social change often creates conflict
    C. social change is a process that affects all segments of society
    D. all of the above

7. The early theorists _____ and Emile Durkheim both argued that as societies move through history, they become more complex.

    A. Herbert Spencer
    B. Talcott Parsons
    C. Todd Gitlin
    D. none of the above

8. Organic solidarity is also called:

    A. homogeneous solidarity
    B. cohesive solidarity
    C. contractual solidarity
    D. mechanical solidarity

9. Functionalist theory is most closely reflected in:

    A. evolutionary theory
    B. world systems theory
    C. dependency theory
    D. feminist theory

10. In support of the overall argument that social change is in fact _____, Lenski and associates point out that many agricultural societies have transformed into industrial societies throughout history.

    A. cyclical
    B. evolutionary
    C. revolutionary
    D. none of the above

11. _____ argued that societies could "advance" and that advancement was to be measured in terms of movement from a class society to one without class.

    A. Gerhard Lenski
    B. Herbert Spencer
    C. Emile Durkheim
    D. Karl Marx

Chapter 23

12. _____ believed that the most important causes of social change were the tensions between various social groups, especially those defined along social class lines.

    A. Gerhard Lenski
    B. Karl Marx
    C. Emile Durkheim
    D. Herbert Spencer

13. The theory which argues that the structural, institutional, and cultural development of a society can follow many evolutionary paths simultaneously, with the different paths all emerging from the circumstances of the society in question is:

    A. multidimensional evolutionary theory
    B. dependency theory
    C. neoevolutionary theory
    D. both A and C

14. Lenski argues that _____ are significantly, although not wholly, responsible for other changes, such as alterations in religious preferences, the nature of law, the form of government, and relations between races and genders.

    A. technological advances
    B. family changes
    C. changes in society's age composition
    D. normative changes

15. Karl Marx was influenced by the early functionalist and evolutionary theories of:

    A. Auguste Comte
    B. Emile Durkheim
    C. Herbert Spencer
    D. none of the above

16. According to sociologist _____, countries where major revolutions have occurred, serious internal conflict between social classes were combined with major international crises that the elite social classes proved unable to resolve before they were overthrown.

    A. Pitrim Sorokin
    B. Theda Skocpol
    C. Theodore Caplow
    D. Gerhard Lenski

17. According to _____, the necessity for growth is the primary cause of social change.

    A. conflict theory
    B. modernization theory
    C. cyclical theory
    D. world systems theory

18. According to _____, the most successful nations control the development of less powerful nations, which become dependent on them.

   A. conflict theory
   B. modernization theory
   C. cyclical theory
   D. dependency theory

19. According to _____, the primary causes of social change are technology and global development.

   A. conflict theory
   B. dependency theory
   C. modernization theory
   D. world systems theory

20. According to cyclical theories of social changes, the third phase of societies, which stresses practical approaches to reality and also involves the hedonistic and the sensual, is called the:

   A. sensate culture
   B. ideational culture
   C. diffused culture x
   D. idealistic culture

21. Which of the following theorists is not associated with a cyclical theory of social change?

   A. Pitrim Sorokin
   B. Arnold Toynbee
   C. Oswald Spengler
   D. Karl Marx

22. _____, a social historian and a principle theorist of social change, argues that societies are born, mature, decay, and sometimes die.

   A. Pitrim Sorokin
   B. Theordore Caplow
   C. Arnold J. Toynbee
   D. Gunnar Myrdal

23. _____ and Theodore Caplow argued that societies proceed through three different phases or cycles of change—idealistic, ideational, sensate.

   A. Oswald Spengler
   B. Pitrim Sorokin
   C. Arnold J. Toynbee
   D. Jeremy Rifkin

Chapter 23

24. The anthropologist _____ alerted us some time ago to the fact that many things people often regard as "American" originally came from other lands.

    A. Horace Miner
    B. Ralph Linton
    C. Oscar Lewis
    D. none of the above

25. According to the text, coins originated in:

    A. Turkey
    B. China
    C. the United States
    D. Peru

26. Researchers have traced *step shows* to traditional:

    A. South Asian religious rituals
    B. South African folklore
    C. West and central African group dances
    D. none of the above

27. *Playing the dozens* and *signifyin'* are not new or unique to the United States. Similar games can be found in certain:

    A. Latin American cultures
    B. Middle Eastern cultures
    C. Eastern European cultures
    D. Pacific Islander cultures

28. The term Ebonics refers to urban:

    A. urban Caribbean English
    B. urban Asian English
    C. urban Black English
    D. urban Latino English

29. Which of the following social theorists presented the culture of calculation and the culture of simulation?

    A. Arnold J. Toynbee
    B. Horace Miner
    C. Sherry Turkle
    D. Oscar Lewis

Social Change in Global Perspective

30. The _____, a nonindustrial agrarian society existing deep in the rain forests of South America, live without electricity, automobiles, guns, and other items of material culture associated with industrialized societies.

   A. Seminole
   B. Maya
   C. Mapuche
   D. Yanomami

31. By the year 2015, _____ percent of the population will be age 55 and older.

   A. 27
   B. 51
   C. 32
   D. 63

32. The _____ is the expectation that the first generation will grow up and raise the second generation, who in turn will produce a third generation. Then, in later life, the younger generations will take care of the older generations.

   A. *filial support contract*
   B. *contract between generations*
   C. *graying together contract*
   D. *generational gap contract*

33. By the year 2050, it is expected that Hispanics will be _____ percent of the U.S. population.

   A. 55
   B. 45
   C. 35
   D. 25

34. _____ is a specific type of social change that industrialization tends to bring about.

   A. Modernization
   B. *Gesellschaft*
   C. *Gemeinschaft*
   D. none of the above

35. _____ refers to a state characterized by a sense of fellow feeling, strong personal ties, and sturdy primary group memberships, along with a sense of personal loyal to one another.

   A. Mass society
   B. *Gesellschaft*
   C. *Gemeinschaft*
   D. none of the above

Chapter 23

36. Another product of modernization is pronounced social stratification, according to:

    A. Robert Merton
    B. Jürgen Habermas
    C. Peter Berger
    D. Ralf Dahrendorf

37. Social theorist _____ argued that three main orientations of personality can be traced to social structural conditions.

    A. Ralf Dahrendorf
    B. Jürgen Habermas
    C. Peter Berger
    D. David Riesman

38. _____ is a structural condition where in the individual is guided by internal principles and morals and is relatively impervious to the superficialities of those around him or her.

    A. Introvert-directedness
    B. Other-directedness
    C. Tradition-directedness
    D. Inner-directedness

39. According to Riesman, modernization tends to produce:

    A. introvert-directedness
    B. inner-directedness
    C. tradition-directedness
    D. other-directedness

40. The influential social theorist _____ has argued that modernized society fails to meet the basic needs of people, among them the need for fulfilling identity.

    A. Ralf Dahrendorf
    B. Jürgen Habermas
    C. Peter Berger
    D. Herbert Marcuse

41. _____ refers to the increased interconnectedness and interdependence of different societies around the world.

    A. Localization
    B. Globalization
    C. Mercantilism
    D. Modernization

42. Modernization theory traces the beginnings of globalization to:

    A. Western Europe
    B. the United States
    C. the former Soviet Bloc
    D. both A and B

43. _____ are cities that themselves connect entire societies.

    A. "International cities"
    B. "Web cities"
    C. "Digital cities"
    D. "World cities"

44. Which of the following is a noncore nation?

    A. Nigeria
    B. India
    C. Brazil
    D. all of the above

45. Which of the following is a core nation?

    A. South Korea
    B. England
    C. Mexico
    D. Greece

46. _____ argues that all nations are members of a worldwide system of unequal political and economic relationships that benefit the developed and technologically advanced countries at the expense of the less technologically advances and less developed countries.

    A. Modernization theory
    B. Functionalist theory
    C. World systems theory
    D. Symbolic Interaction theory

47. According to _____ theory, industrialized nations imprison developing nations in dependent relationships.

    A. modernization
    B. cyclical
    C. dependency
    D. social exchange

Chapter 23

48. Former Secretary of _____ Robert Reich has noted that core nations have been willing to lend money to noncore nations, but often on terms such as high interest rates that put severe economic strain on the noncore nations.

   A. Defense
   B. Homeland Security
   C. Labor
   D. Education

49. As economist Reich and sociologists McCord and McCord have noted, the network of dependency is complicated by the fact that in today's global economic system, it is difficult to determine:

   A. just who owns what
   B. the impact on individual nations
   C. who started it all
   D. none of the above

50. _____ as a cause of change is exemplified by the effects of immigration into a country and the resultant changes in society.

   A. Predictability
   B. Diversity
   C. Standardization
   D. Homogeneity

True/False Questions:

T F 1. Macrochanges are subtle alterations in the day to day interaction between people.

T F 2. Social change often creates conflict and is not random.

T F 3. Evolutionary theories of social change are a branch of conflict theory.

T F 4. A central theme for Durkheim is that revolution and dramatic change would come about when class conflict led inevitably to a decisive social rupture.

T F 5. Cyclical theories of social change build on the idea of economic conflict between social classes.

T F 6. Arnold J. Toynbee is the principal theorist of cyclical social change.

T F 7. Cultural diffusion is the transmission of cultural elements from one society or cultural group to another.

T F 8. Culture in and of itself cannot contribute to the persistence of social inequality and become a source of discontent among the individuals in the society.

Social Change in Global Perspective

T   F   9.   The advent of the electronic computer has massively transformed our entire society and all of its institutions.

T   F   10.   By the year 2025 the Hispanic population will account for 18 percent of the U.S. population.

T   F   11.   The concept of mass society is one in which there will be unified primary, family, and kinship ties.

T   F   12.   Marx argued that the capitalist system will eventually overcome the inequalities of society.

T   F   13.   Futurist Marshall McLuhan called the increasingly interconnected, homogeneous culture a "global village."

T   F   14.   Greater interconnectedness among societies may magnify the cultural differences between interacting groups.

T   F   15.   Modernization theory states that all nations are members of a worldwide system of unequal political and economic relationships.

T   F   16.   Wallerstein's world systems theory divides into two camps: core and noncore nations.

T   F   17.   There is an economic dependency of noncore nations upon the core nations.

T   F   18.   Dependency theory maintains that highly industrialized nations will come to the aid of Third World countries and develop them into thriving capitalist systems.

T   F   19.   Social change can affect the relations between societies.

T   F   20.   The effects of technology and modernization have defeated the barriers of diversity on a global scale.

Short Answer Questions:

1. Compare and contrast between microchanges and macrochanges. Provide examples.
2. What is cultural diffusion? Provide examples.
3. List and discuss the three general characteristics of modernization.
4. Compare and contract between other-directedness and inner-directedness.

Chapter 23

ANSWERS FOR CHAPTER TWENTY-THREE

Multiple Choice Questions:

1. D (p.618)
2. C (p.618)
3. D (p.618)
4. B (p.619)
5. B (p.619)
6. D (pp.619-620)
7. A (p.621)
8. C (p.621)
9. A (p.621)
10. B (p.622)
11. D (p.622)
12. B (p.622)
13. A (p.622)
14. A (p.622)
15. C (p.622)
16. B (p.623)
17. C (p.624, Table 23.1)
18. D (p.624, Table 23.1)
19. C (p.624, Table 23.1)
20. A (p.624)
21. D (p.624)
22. C (p.624)
23. B (p.624)
24. B (p.625)
25. A (p.625)
26. C (p.626)
27. B (p.626)
28. C (p.627)
29. C (p.628)
30. D (p.628)
31. A (p.630)
32. B (p.630)
33. D (p.630)
34. A (p.631)
35. C (p.631)
36. B (p.632)
37. D (p.633)
38. D (p.633)
39. D (p.633)
40. D (p.633)
41. B (p.634)
42. D (p.634)
43. D (p.634)
44. D (p.634)
45. B (p.634)
46. C (p.634)
47. C (p.635)
48. C (p.635)
49. A (p.635)

50. B (p.636)

True/False Questions:

1. F (p.618)
2. T (p.620)
3. F (p.621)
4. F (p.621)
5. F (p.624)
6. T (p.624)
7. T (p.625)
8. F (p.627)
9. T (p.628)
10. T (p.630)
11. F (p.632)
12. F (p.632)
13. T (p.634)
14. T (p.634)
15. F (p.634)
16. T (p.634)
17. T (p.634)
18. F (p.635)
19. T (p.636)
20. F (p.636)

Short Answer Questions:

1. See p.618
2. See p.625
3. See p.631
4. See p.633